I0764015

FOURTEEN WEEKS

IN

DESCRIPTIVE ASTRONOMY.

BY

J. DORMAN STEELE, PH.D.

AUTHOR OF THE FOURTEEN-WEEKS SERIES IN PHYSIOLOGY, PHILOSOPHY, CHEMISTRY, AND GEOLOGY.

"The heavens declare the glory of God; and the firmament showeth his handy-work."
PSALM xix. 1.

A. S. BARNES & COMPANY,

NEW YORK AND CHICAGO.

1874.

THE FOURTEEN WEEKS' COURSES

IN

NATURAL SCIENCE,

BY

J. DORMAN STEELE, A.M., PH.D.

Fourteen Weeks in Natural Philosophy, . . $1.50
Fourteen Weeks in Chemistry, ... 1.50
Fourteen Weeks in Descriptive Astronomy, . 1.50
Fourteen Weeks in Popular Geology, . . 1.50
Fourteen Weeks in Human Physiology, . . 1.50
A Key, containing Answers to the Questions and Problems in Steele's 14 Weeks' Courses, 1.50

A HISTORICAL SERIES,

on the plan of Steele's 14 Weeks in the Sciences,

inaugurated by

A Brief History of the United States, . . 1.50

The publishers of this volume will send either of the above by mail, post-paid, on receipt of the price.

The same publishers also offer the following standard scientific works, being more extended or difficult treatises than those of Prof. Steele, though still of Academic grade.

Peck's Ganot's Natural Philosophy, . . . 1.75
Porter's Principles of Chemistry, ... 2.00
Jarvis' Physiology and Laws of Health, . 1.65
Wood's Botanist and Florist, ... 2.50
Chambers' Elements of Zoology, ... 1.50
McIntyre's Astronomy and the Globes, . . 1.50
Page's Elements of Geology, ... 1.25

Address A. S. BARNES & CO.,
Educational Publishers,
NEW YORK OR CHICAGO.

STEELE'S AST.

PREFACE.

During the past few years great advances have been made in astronomical science. A new horizontal parallax of the sun has been established. This has materially altered the estimated distances, etc., of the planets. The sun is much nearer us than we supposed, and light has lost a little of its wonderful velocity. Much additional information has been obtained concerning Meteors and Shooting Stars. The investigations connected with Spectrum Analysis have been especially suggestive. Thus on every hand the facts of Astronomy have been accumulating. As yet, however, they are scattered through many expensive foreign works, and are consequently beyond the reach of most of our schools. It has been the aim of the author to collect in this little volume the most interesting features of these larger works. Believing that Natural Science is full of fascination, he has sought to weave the story of those far-distant worlds into a form that may attract the attention and kindle the enthusiasm of the pupil. The work is not written for the information of scientific men, but for the inspiration of youth. The pages therefore are not burdened with a multitude of figures which no memory could possibly retain. Mathematical tables and data, Questions for Review, and also a Guide to the Constellations, are given in the Appendix, where they may be useful for constant reference.

The author would call particular attention to the method of classifying the measurements of Space, and the practical treatment of the subjects of Parallax, Harvest Moon, Eclipses, the Seasons, Phases of the Moon, Time, Nebular Hypothesis, &c.

To teachers heretofore compelled to use a cumbersome set of charts, it is hoped that the star maps here offered will present a welcome substitute. The geometrical figures showing the actual appearance of the constellations, will relieve the mind confused with the idea of numberless rivers, serpents, and classical heroes. The brightest stars only are given, since in practice it is found that pupils remember the general outlines alone.

Finally, the author commits this little work to the hands of the young, to whose instruction he has consecrated the energies of his life, in the earnest hope that, loving Nature in all her varied phases, they may come to understand somewhat of the wisdom, power, beneficence, and grandeur displayed in the Divine harmony of the Universe.

> "One God, one law, one element,
> And one far-off Divine event
> To which the whole creation moves."

The following works, among others, have been freely consulted in preparing this volume:

The Heavens.....................Guillemin.
Astronomy.......................Chambers.
Introduction to Astronomy.......Hind.
Solar System....................Hind.
Popular Astronomy...............Airy.
Popular Astronomy...............Arago.
Astronomy.......................Norton.
Astronomy.......................Robinson.
Astronomy.......................Loomis.
Age of Fable....................Bulfinch.
Poetry of Science...............Hunt.
Outlines of Astronomy...........Herschel.
Popular Astronomy...............Mitchell.
Astronomy and Physics...........Whewell.
Annual of Scientific Discovery.....Kneeland.
The Chemical News.

Publishers' Notice.—Teachers will find in each edition of this Series certain changes; not such, however, as to cause any inconvenience in the use of all the editions in the same class. These are to be considered, not as corrections of errors, but as improvements suggested by the constant advance made in science, and by practical work in the school-room. The publishers are determined to spare no expense in making this Series increasingly worthy of the unprecedented success it has attained.

SUGGESTIONS TO TEACHERS.

This work is designed to be recited in the topical method. On naming the title of a paragraph, the pupil should be able to draw on the blackboard the diagram, if any is given, and state the substance of what is contained in the book. It will be noticed that the order of topics, in treating of the planets and also of the constellations, is uniform. If a portion of the class write their topics in full upon the blackboard, it will be found a valuable exercise in spelling, punctuation, and composition. Every point which can be illustrated in the heavens should be shown to the class. No description or apparatus can equal the reality in the sky. After a constellation has been traced, the pupil should be practised in star-map drawing. Much profitable instruction can be obtained in this way. For the purpose of more easily finding the heavenly bodies at any time, Whitall's Movable Planisphere is of great service. It may be procured of the publishers of this work. "Orreries are of little account." A tellurian is invaluable in explaining Precession of the Equinoxes, Eclipses, Phases of the Moon, etc. Messrs. A. S. Barnes & Co., New York City, furnish a good instrument at a low price. The article on "Celestial Measurements," near the close of the work, should be constantly referred to during the term. In the figures, the right-hand side represents the west and the left-hand the east. When it is important to obtain this idea correctly, the book should be held up toward the southern sky.

Never let a pupil recite a lesson, nor answer a question, except it be a mere definition, *in the language of the book.* The text is designed to interest and instruct the pupil; the recitation should afford him an opportunity of expressing what he has learned, in his own style and words.

TABLE OF CONTENTS.

INTRODUCTION.

Astronomy (*astron*, a star, and *nomos*, a law) treats of the Heavenly Bodies—the sun, moon, planets, stars, and, as our globe itself is a planet, of the earth also. It is, above all others, a science that cultivates the powers of the imagination. Yet all its theories and distances are based upon the most rigorous mathematical demonstrations. Thus the study has at once the beauty of poetry and the exactness of Geometry.

The Appearance of the Heavens to an Observer.—The great dome of the sky filled with glittering stars is one of the most sublime spectacles in nature. To enjoy this fully, a night must be chosen when the air is clear, and the moon is absent. We then gaze upon a deep blue, an immense expanse studded with stars of varied color and brilliancy. Some shine with a vivid light, perpetually changing and twinkling; others, more constant, beam tranquilly and softly upon us; while many just tremble into our sight, like a wave that, struggling to reach some far-off land, dies as it touches the shore. In the presence of such weird and wondrous beauty, the tenderest sentiments of the heart are aroused—a feeling of awe and reverence, of softened melancholy mingled with a thought of God, comes over us, and awakens the better nature within us. Those far-off lights seem full of meaning to us, could we but read their holy message ; they become real and sentient, and, like the soft eyes in pictures, look lovingly and inquiringly upon us. We come into communion with another life, and the soul asserts its immortality more strongly than ever before. We are humbled as we gaze upon the infinity of worlds, and strive to comprehend their enormous distances, their magnificent retinue of suns. The powers of the mind are aroused, and eager questionings crowd upon us. What are those glittering fires? What their distances from us? Are they worlds like our own? Do living, thinking beings dwell upon them? Are they carelessly scattered through infinite space, or is there an order of the universe? Can we ever hope to fathom those mysterious depths, or are they closed to us forever? Many of these problems have been solved; others yet await the astronomer whose keen eye shall be strong enough to read the mysterious scroll of the heavens. Two hundred generations of study have revealed to us such startling facts, that we wonder how man in his feebleness can grasp so much, see so far, and penetrate so deeply into the mysteries of the universe. Astronomy has measured the distance of many of the stars, and of all the planets; computed their weight and size, their days, years, and seasons, with many of their physical features; made a map of the moon, in some respects more perfect than any map of the earth; tracked the comets in their immense sidereal journeys, marking their paths to a nicety of which we can scarcely conceive, and at last it has analyzed the structure of the sun and far-off stars, announcing the very elements of which they are composed.

Observing for several evenings those stars which shine with a clear distinct light, we notice that they change their position with respect to the others. They are therefore called "*planets*" (literally, *wanderers*). Others remain immovable, and shine with a shifting, twinkling light. They are termed the "*fixed stars*," although it is now known that they also are in motion—the most rapid of any known even to Astronomy—but through such immense orbits that they seem to us stationary. Then, too, diagonally girdling the heavens, is a whitish, vapory belt—the *Milky Way*. This is composed of multitudes of millions of suns—of which our own sun itself is one—so far removed from us that their light mingles, and makes only a fleecy whiteness. This magnificent panorama of the heavens is before us, inviting our study, and waiting to make known to us the grandest revelations of science.

DESCRIPTIVE ASTRONOMY.

HISTORY.

Astronomy is the most ancient of all sciences. The study of the stars is doubtless as old as man himself, and hence many of its discoveries date back of authentic records, amid the dim mysteries of tradition. In tracing its history, we shall speak only of those prominent facts which will best enable us to understand its progress and glorious achievements.

The Chinese.—This people boast much of their astronomical discoveries. Indeed their emperor claims a celestial ancestry, and styles himself "Son of the Sun." They possess an account of a conjunction of four planets and the moon, which must have occurred a century before the Flood. They have also the first record of an eclipse of the sun, which took place about two hundred and twenty years* after the Deluge. It is reported that one of their kings, two thousand years before Christ, put to death the principal officers of state because they had failed to calculate an approaching eclipse.

* October 13, 2127 B. C.

THE CHALDEANS.—The Chaldean shepherds, watching their flocks by night under the open sky, could not fail to become familiar with many of the movements of the heavenly bodies. When Alexander took Babylon, two centuries before Christ, he found in that city a record of their observations reaching back about nineteen centuries, or nearly to the confusion of tongues at the Tower of Babel. The Chaldeans divided the day into twelve hours, invented the sun-dial, and also discovered the "Saros" or "Chaldean Period," which is the length of time in which the eclipses of the sun and moon repeat themselves in the same order.

THE GRECIANS.—In the seventh century B. C., *Thales*, noted for his electrical discoveries, acquired much renown, and established the first school of Astronomy in Greece. He taught that the earth is round, and that the moon receives her light from the sun. He introduced the division of the earth's surface into zones, and the theory of the obliquity of the ecliptic. He also predicted an eclipse of the sun which is memorable in ancient history as having terminated a war between the Medes and Lydians. These nations were engaged in a fierce battle, but the awe produced by the darkening of the sun was so great, that both sides threw down their arms and made peace. Thales had two pupils, *Anaximander* and *Anaxagoras*. The first of these taught that the stars are suns, and that the planets are inhabited. He erected the first sun-dial, at Sparta. The second maintained that there is but one God, that the sun is solid, and as large as the country of Greece, and attempted to explain eclipses and other celestial phenomena by natural causes. For his audacity and impiety, as his countryman considered it, he and his family were doomed to perpetual banishment.

Pythagoras founded the second celebrated astronomical school, at Crotona, at which were educated hundreds of enthusiastic pupils. He knew the causes of eclipses, and calculated them by means of the Saros. He was most emphatically a dreamer. He conceived a system of the universe, in many respects correct; yet he advanced no proof, and made few converts to his views, and they were soon well-nigh forgotten. He held that the sun is the centre of the solar system, and that the planets revolve about it in circular orbits; that the earth revolves daily on its axis, and yearly around the sun; that Venus is both morning and evening star; that the planets are inhabited—and he even attempted to calculate the size of some of the animals in the moon; that the planets are placed at intervals corresponding to the scale in music, and that they move in harmony, making the "music of the spheres," but that this celestial concert is heard only by the gods—the ears of man being too gross for such divine melody.

Eudoxus, who lived in the fourth century B. C., invented the theory of the Crystalline Spheres. He held that the heavenly bodies are set, like gems, in hollow, transparent, crystal globes, which are so pure that they do not obstruct our view, while they all revolve around the earth. The planets are placed in one globe, but have a power of moving themselves, under the guidance—as Aristotle taught—of a tutelary genius, who resides in each, and rules over it as the mind rules over the body.

Hipparchus, who flourished in the second century B. C., has been called the "Newton of Antiquity." He was the most celebrated of the Greek astronomers. He calculated the length of the year to within six minutes, discovered the precession of the equinoxes, and made the first catalogue of the stars—1081 in number.

THE EGYPTIANS.—Egypt, as well as Chaldea, was noted for its knowledge of the sciences long before they were cultivated in Greece. It was the practice of the Greek philosophers, before aspiring to the rank of teacher, to travel for years through these countries, and gather wisdom at its fountain-head. Pythagoras spent thirty years in this manner. Two hundred years after Pythagoras, the celebrated school of Alexandria was established. Here were concentrated in vast libraries and princely halls nearly all the wisdom and learning of the world. Here flourished all the sciences and arts, under the patronage of munificent kings. At this school Ptolemy, a Grecian, wrote his great work, the "Almagest," which for fourteen centuries was the textbook of astronomers. In this work was given what is known as the "Ptolemaic System." It was founded largely upon the materials gathered by previous astronomers, such as Hipparchus, whom we have already mentioned, and Eratosthenes, who computed the size of the earth by the means even now considered the best—the measurement of an arc of the meridian.

PTOLEMAIC THEORY.—The movements of the planets were to the ancients extremely complex. Venus, for instance, was sometimes seen as "evening star" in the west, and then again as "morning star" in the east. Sometimes she seemed to be moving in the same direction as the sun, then going apparently behind the sun, appeared to pass on again in a course directly opposite. At one time she would recede from the sun more and more slowly and coyly, until she would appear to be entirely stationary; then she would retrace her steps, and seem to meet the sun. All these facts were attempted to be accounted for by an incongruous system of "cycles and epicycles," as it is called. The advocates of this theory assumed that every planet revolves in a circle, and that the earth is the fixed centre around which the sun and the heavenly bodies move. They then conceived that a bar, or something equivalent, is connected at one end with the earth; that at some part of this bar the sun is attached; while between that and the earth, Venus is fastened—not to the bar directly, but to a sort of crank; and further on, Mercury is hitched on in the same way. In the cut, let A be the earth, S the sun, ABDF the bar (real or imaginary), BC the short bar or crank to which Venus is tied, DE another bar for Mercury, FG another bar, with still another short crank, at the end of which, H, Mars is attached.

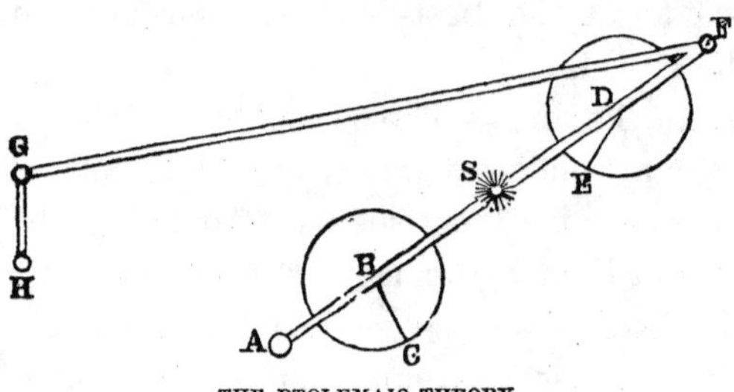

THE PTOLEMAIC THEORY.

Thus they had a complete system. They did not exactly understand the nature of these bars—whether they were real or only imaginary—but they *did* comprehend their action, as they thought; and so they supposed the bar revolved, carrying the sun and planets along in a large circle about the earth; while all the short cranks kept flying around, thus sweeping each planet through a smaller circle. By this theory, we can see that the planets would sometimes go in front of the sun and sometimes behind; and their places were so accurately predicted, that the error could not be detected by the rude instruments then in use. As soon as a new motion of one of the heavenly bodies was discovered, a new crank, and of course a new circle, was added to account for the fact. Thus the system became more and more complicated, until a combination of five cranks and circles was necessary to make the planet Mars keep pace with the Ptolemaic theory. No wonder that Alfonso, king of Castile, and a very celebrated patron of Astronomy, revolted at the cumbersome machinery, and cried out, "If I had been consulted at the creation, I could have done the thing better than that!"

ASTROLOGY.—After the death of Ptolemy, Astronomy ceased to be cultivated as a science. The Romans, engrossed with schemes of conquest, never produced a single great astronomer. Indeed, when Julius Cæsar reformed the calendar, he obtained the assistance, not of a Roman, but of Sosigenes an Alexandrian. The Arabians studied the stars merely for purposes of soothsaying and prophecy. They professed to foretell the future by the appearance of the planets or stars. All of the ancient astronomers shared more or less in this superstition. Tiberius, emperor of Rome, practised Astrology. Hippocrates himself, the "Father of Medicine" who flourished in the 4th century B. C., ranked it among the most important branches of knowledge for the physician. Star-diviners were held in the greatest estimation. The system continued to increase in credit until the Middle Ages, when it was at its height of popularity. The issue of any important undertaking, or the fortunes of an individual, were foretold by the astrologer, who drew up a *Horoscope*, representing the position of the stars and planets at the beginning of the enterprise, or at the birth of the person. It was a complete and complicated system, and contained regular rules, which guided the interpretation, and which were so abstruse that they required years for their entire mastery. Venus foretold love; Mars, war; the Pleiades,* storms at sea. The ignorant were not alone the dupes of this visionary system. Lord Bacon believed in it most firmly. As late even as the reign of Charles II., Lilly, a famous astrologer of that time, was called before a committee of the House of Commons to give his opinion on the probable issue of some enterprise then under consideration. However foolish the system of Astrology itself may have been, it preserved the science of Astronomy during the Dark Ages, and prompted to accurate observation and diligent study of the heavens.

The Copernican System.—About the middle of the sixteenth century, Copernicus, breaking away from the theory of Ptolemy, which was still taught in all the institutions of learning in Europe, revived the theory of Pythagoras. He saw how beautifully simple is the idea of considering the sun the grand centre about which revolve the earth and all the planets. He noticed how constantly, when we are riding swiftly, we forget our own motion, and think that the trees and fences are gliding by us in

* Plē′-ya-dēz.

the contrary direction. He applied this thought to the movements of the heavenly bodies, and maintained that, instead of all the starry host revolving about the earth once in twenty-four hours, the earth simply turns on its own axis: that this produces the apparent daily revolution of the sun and stars; while the yearly motion of the earth about the sun, transferred in the same manner to that body, would account for its various movements. Though Copernicus thus simplified so greatly the Ptolemaic theory, he yet found that the idea of circular orbits for the planets would not explain all the phenomena; he therefore still retained the "cycles and epicycles" that Alfonso had so heartily condemned. For forty years this illustrious astronomer carried on his observations in the upper part of a humble, dilapidated farm-house, through the roof of which he had an unobstructed view of the sky. The work containing his theory was at last published just in time to be laid upon his death-bed.

TYCHO BRAHÉ, a celebrated Danish astronomer, next propounded a modification of the Copernican system. He rejected the idea of cycles and epicycles, but, influenced by certain passages of Scripture, maintained, with Ptolemy, that the earth is the centre, and that all the heavenly bodies revolve about it daily in circular orbits. Brahé was a nobleman of wealth, and, in addition, received large sums from the Government. He erected a magnificent observatory, and made many beautiful and rare instruments. Clad in his robes of state, he watched the heavens with the intelligence of a philosopher and the splendor of a king. His indefatigable industry and zeal resulted in the accumulation of a vast fund of astronomical knowledge, which, however, he lacked the wit to apply to any further advance in science. His pupil, Kepler, saw these facts, and in his fruitful mind they germinated into three great truths, called Kepler's laws. These constitute almost the sum of astronomical knowledge, and form one of the most precious conquests of the human mind. They are the three arches of the bridge over which Astronomy crossed the gulf between the Ptolemaic and Copernican systems.

KEPLER'S LAWS.—Kepler, taking the investigations of his master, Tycho Brahé, determined to find what is the exact shape of the orbits of the planets. He adopted the Copernican theory, that the sun is the centre of the system. At that time all believed the orbits to be circular. Since, as they said, the circle is perfect, is the most beautiful figure in nature, has neither beginning nor ending, therefore it is the only form worthy of God, and He must have used it for the orbits of the worlds He has made. Imbued with this romantic view, Kepler commenced with a rigorous comparison of the places of the planet Mars, as observed by Brahé, with the places as stated by the best tables that could be computed on the circular theory. For a time they agreed, but in certain portions of the orbit the observations of Brahé would not fit the computed place by eight minutes of a degree. Believing that so good an astronomer could not be mistaken as to the facts, Kepler exclaimed, "Out of these eight minutes we will construct a new theory that will explain the movements of all planets." He resumed his work, and for eight years continued to imagine every conceivable hypothesis, and then patiently to test it—"hunt it down," as he called it. Each in turn proved false, until nineteen had been tried. He then determined to abandon the circle and adopt another form. The *ellipse* suggested itself to his mind. Let us see how this figure is made.

Fig. 2.

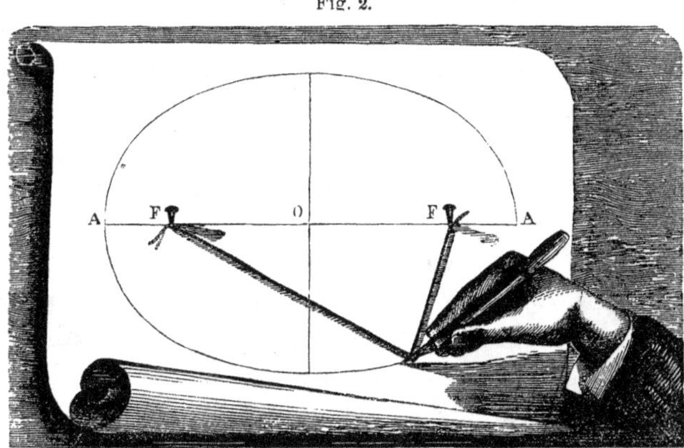

Attach a thread to two pins, as at F F in the figure; next move a pencil along with the thread, the latter being kept tightly stretched, and the point will mark a curve which is flattened in proportion to the length of the string we use,—the longer the string, the nearer a circle will the figure become. This figure is the *ellipse.* The two points F F are called the *foci* (singular, *focus*). We can now understand Kepler's attempt, and the glorious triumph which crowned his seventeen years of unflagging toil.

First Law.—With this figure he constructed an orbit, having the sun at the centre, and again followed the planet Mars in its course. But very soon there was as great discrepancy between the observed and computed places as before. Undismayed by this failure, Kepler assumed another hypothesis. He determined to place the sun at one of the foci of the ellipse, and once more "hunted down" the theory. For a whole year he traced the planet along the imaginary orbit, and it did not diverge. The truth was discovered at last, and Kepler announced his first great law—

PLANETS REVOLVE IN ELLIPSES, WITH THE SUN AT ONE FOCUS.

Second Law.—Kepler knew that the planets do not move with equal velocity in the different parts of their orbits. He next set about establishing some law by which this speed could be determined, and the place of the planet computed. He drew an ellipse, and marked the various positions of the planet Mars once more. He soon found that when at its *perihelion* (point nearest the sun) it moves the fastest, but when at its *aphelion* (point furthest from the sun) it moves the slowest. Once more he "hunted down" various hypotheses, until at last he discovered that while in going from B to A the planet moves very slowly, and from D to C

Fig. 3.

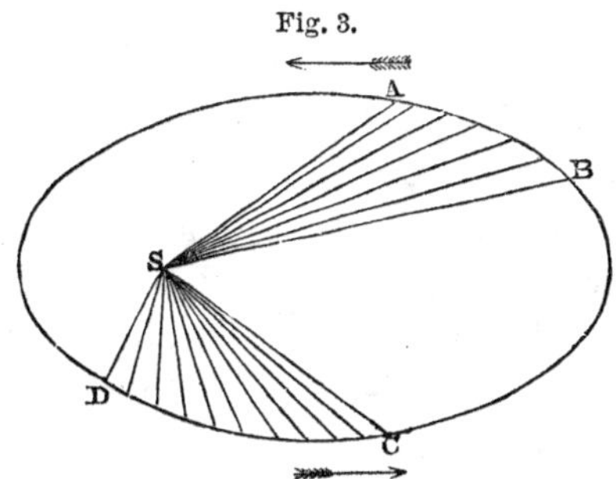

very rapidly; yet the space inclosed between the lines SB and SA is equal to that inclosed between SD and SC. Hence the second law—

A LINE CONNECTING THE CENTRE OF THE EARTH WITH THE CENTRE OF THE SUN, PASSES OVER EQUAL SPACES IN EQUAL TIMES.

Third Law.—Kepler, not satisfied with the discovery of these laws, now determined to ascertain if there were not some relation existing between the times of the revolution of the planets about the sun and their distances from that body. With the same wonderful patience, he took the figures of Tycho Brahé, and began to compare them. He tried them in every imaginable relation. Next he took their squares, then he attempted their cubes, and lastly he combined the squares and the cubes. Here was the secret; but he toiled around it, made a blunder, and waited for months, until, once more, his patience triumphed, and he reached the third law—

THE SQUARES OF THE TIMES OF REVOLUTION OF THE PLANETS ABOUT THE SUN, ARE PROPORTIONAL TO THE CUBES OF THEIR MEAN DISTANCES FROM THE SUN.*

In rapture over the discovery of these three laws, so marked by that divine simplicity which pervades all the laws of nature, Kepler exclaimed, "Nothing holds me. The die is cast. The book is written, to be read now or by posterity, I care not which. It may well wait a century for a reader, since God has waited six thousand years for an observer."†

Galileo.—Contemporary with Kepler was the great Florentine philosopher, Galileo. He discovered the laws of the pendulum and of falling bodies, as we have already learned in Natural Philosophy. He, however, was educated in and believed the Ptolemaic theory. A disciple of the Copernican theory happening to come to Pisa, where Galileo was teaching

---

* For example: The square of Jupiter's period is to the square of Mars' period, as the cube of Jupiter's distance is to the cube of Mars' distance; or, representing the earth's time of revolution by P, and her distance from the sun by p, then letting D and d represent the same in another planet, we have the proportion $P^2 : D^2 :: p^3 : d^3$.

† Kepler, strangely enough, believed in the "Music of the Spheres." He made Saturn and Jupiter take the bass, Mars the tenor, Earth and Venus the counter, and Mercury the treble. This shows what a streak of folly or superstition may run through the character of the noblest man. However, as Johnson says, a mass of metal may be gold, though there be in it a little vein of tin.

as professor in the University, drew his attention to its simplicity and beauty. His clear discriminating mind perceived its perfection, and he henceforth advocated it with all the ardor of his unconquerable zeal. Soon after he learned that one Jansen, a Dutch watchmaker, had invented a contrivance for making distant objects appear near. With his profound knowledge of optics and philosophical instruments, Galileo instantly caught the idea, and soon had a telescope completed that would magnify thirty times. It was a very simple affair—only a piece of lead pipe with glasses set at each end; but it was the first telescope ever made, and destined to overthrow the old Ptolemaic theory, and revolutionize the whole science of Astronomy.

Discoveries made with the telescope.—Galileo now examined the moon. He saw its mountains and valleys, and watched the dense shadows sweep over its plains. On January 8, 1610, he turned the telescope toward Jupiter. Near it he saw three bright stars, as he considered them, which were invisible to the naked eye. The next night he noticed that those stars had changed their relative positions. Astonished and perplexed, he waited three days for a fair night in which to resume his observations. The fourth night was favorable, and he again found the three stars had shifted. Night after night he watched them, discovered a fourth star, and finally found that they were all rapidly revolving around Jupiter, each in its elliptical orbit, with its own rate of motion, and all accompanying the planet in its journey around the sun. Here was a miniature Copernican system, hung up in the sky for all to see and examine for themselves.

Reception of the discoveries.—Galileo met with the most bitter opposition. Many refused to look through the telescope lest they might become victims of the philosopher's magic. Some prated of the wickedness of digging out valleys in the fair face of the moon. Others doggedly clung to the theory they had held from their youth up. As a specimen of the arguments adduced against the new system, the following by Sizzi is a fair instance. "There are seven windows in the head, through which the air is admitted to the body, to enlighten, to warm, and to nourish it,—two nostrils, two eyes, two ears, and one mouth. So in the heavens there are two favorable stars, Jupiter and Venus; two unpropitious, Mars and Saturn; two luminaries, the Sun and Moon; and Mercury alone, undecided and indifferent. From which, and from many other phenomena of Nature, such as the seven metals, etc., we gather that the number of planets is necessarily seven. Moreover, the satellites are invisible to the naked eye, can exercise no influence over the earth, and would be useless, and therefore do not exist. Besides, the week is divided into seven days, which are named from the seven planets. Now, if we increase the number of planets, this whole system falls to the ground."

NEWTON.—As we have seen, the truth of the Copernican system was fully established by the discoveries of Galileo with his telescope. Philosophers gradually adopted this view, and the Ptolemaic theory became a relic of the past. In 1666, Newton, a young man of twenty-four years, was spending a season in the country, on account of the plague which prevailed at Cambridge, his place of residence. One day, while sitting in a garden, an apple chanced to fall to the ground near him. Reflecting upon the strange power that causes all bodies thus to descend to the earth, and remembering that this force continues, even when we ascend to the tops of high mountains, the thought occurred to his mind, "May not this same force extend to a great distance out in space? Does it not reach the moon?"

Laws of Motion.—To understand the philosophy of the reasoning that now occupied the mind of Newton, let us apply the laws of motion as we have learned them in Philosophy. When a body is once set in motion, it will continue to move forever in a straight line, unless another force is applied. As there is no friction in space, the planets do not lose any of their original velocity, but move now with the same speed which they received in the beginning from the Divine hand. But this would make them all pass through straight, and not circular orbits. What causes the curve? Obviously another force. For example: I throw a stone into the air. It moves not in a straight line, but in a curve, because the earth constantly bends it downward.

Application.—Just so the moon is moving around the earth, not in a straight line, but in a curve. Can it not be that the earth bends *it* downward, just as it does the stone? Newton knew that a stone falls toward the earth sixteen feet the first second. He imagined, after a careful study of Kepler's laws, that the attraction of the earth diminishes according to the square of the distance. He knew (according to the measurement then received) that a body on the surface of the earth is four thousand miles from the centre. He applied this imaginary law. Suppose it is removed four thousand miles from the surface of the earth, or eight thousand miles from the centre. Then, as it is twice as far from the centre, its weight will be diminished 2^2, or 4 times. If it were placed 3, 4, 5, 10 times further away, its weight would then decrease 9, 16, 25, 100 times. If, then, the stone at the surface of the earth (four thousand miles from the centre) falls sixteen feet the first second, at eight thousand miles it would fall only four feet; at 240,000 miles, or the distance of the moon, it would fall only about one-twentieth of an inch (exactly .053). Now the question arose, "How far does the moon fall toward the earth, *i. e.*, bend from a straight line, every second?" For seventeen years, with a patience rivalling Kepler's, this philosopher toiled over interminable columns of figures to find how much the moon's path around the earth curves each second. He reached the result at last. It was nearly, but not quite exact. Disappointed, he laid aside his calculations. Repeatedly he reviewed them, but could not find a mistake. At length, while in London, he learned of a new and more accurate measurement of the distance from the circumference to the centre of the earth. He hastened home, inserted this new value in his calculations, and soon found that the result would be correct. Overpowered by the thought of the grand truth just before him, his hand faltered, and he called upon a friend to complete the computation.

From the moon, Newton passed on to the other heavenly bodies, calculating and testing their orbits. At last he turned his attention to the sun, and, by reasoning equally conclusive, proved that the attraction of that great central orb compels all the planets to revolve about it in elliptical orbits, and holds them with an irresistible power in their appointed paths. At last he announced this grand *Law of Gravitation:*

EVERY PARTICLE OF MATTER IN THE UNIVERSE ATTRACTS EVERY OTHER PARTICLE OF MATTER WITH A FORCE DIRECTLY PROPORTIONAL TO ITS QUANTITY OF MATTER, AND DECREASING AS THE SQUARE OF THE DISTANCE INCREASES.

SPACE.

We now in imagination pass into space, which stretches out in every direction without bounds or measures. We look up to the heavens and try to locate some object among the mazes of the stars. We are bewildered, and immediately feel the necessity of some system of measurement. Let us try to understand the one adopted by astronomers.

THE CELESTIAL SPHERE.—The blue arch of the sky, as it appears to be spread above us, is termed the Celestial Sphere. There are two points to be noticed here. *First*, that so far distant is this imaginary arch from us, that if any two parallel lines from different parts of the earth are drawn to this sphere, they will apparently intersect. Of course this cannot be the fact; but the distance is so immense, that we are unable to distinguish the little difference of four or even eight thousand miles, and the two lines will seem to unite: so we must consider this great earth as a mere speck or point at the centre of the Celestial Sphere. *Second*, that we must even neglect the entire diameter of the earth's orbit, so that if we should draw two parallel lines, one from each end of the earth's orbit, to the sphere, although these lines would be 183,000,000 miles apart, yet they would be extended so far that we could not separate them, and they would appear to pierce the sphere at the same point; which is to say, that at that enormous distance, 183,000,000 miles shrink to a point. Consequently, in all parts of the earth, and in every part of the earth's orbit, we see the fixed stars in the same place. This sphere of stars surrounds the earth on every side. In the daytime we cannot see the stars because of the superior light of the sun; but with a telescope they can be traced, and an astronomer will find certain stars as well at noon as at midnight. Indeed, when looking at the sky from the bottom of a deep well or lofty chimney, if a bright star happens to be directly overhead, it can be seen with the naked eye even at midday. In this way it is said a celebrated optician was first led to think of there being stars by day as well as by night. One half of the sphere is constantly visible to us; and so far distant are the stars, that we see just as much of the sphere as we would if the upper part of the earth were removed, and we were to stand four thousand miles further away, or at the very centre of the earth, where our view would be bounded by a great circle of the earth. On the concave surface of the celestial sphere there are imagined to be drawn *three systems of circles:* the HORIZON, the EQUINOCTIAL, and the ECLIPTIC Systems. Each of these has (1) its *Principal Circle,* (2) its *Subordinate Circles,* (3) its *Points,* and (4) its *Measurements.*

I. The Horizon System.

(*a*) The Principal Circle is the *Rational Horizon.* This is the great circle that, passing through the centre of the earth, separates the visible from the invisible heavens. The *Sensible Horizon* is the small circle where the earth and sky seem to meet; it is parallel to the rational horizon, but distant from it the semi-diameter of the earth. No two places have the same sensible horizon: any two on opposite sides of the earth have the same rational horizon.

(*b*) The Subordinate Circles.—These are the *Prime Vertical* circle and the *Meridian.* A vertical circle is one passing through the poles of the horizon (the zenith and nadir). The Prime Vertical is a vertical circle passing through the East and West points. The Meridian is a vertical circle passing through the North and South points.

(*c*) Points.—These are the *Zenith,* the *Nadir,* the N., S., E., and W. points. The Zenith is the point directly overhead, and the Nadir the one directly underfoot. They are also the poles of the horizon —*i. e.,* the points where the axis of the horizon pierces the celestial sphere. The N., S., E., and W. points are familiar to all.

(*d*) Measurements.—These are *Azimuth, Amplitude, Altitude,* and *Zenith distance.*

Azimuth is the distance from the meridian, measured East or West, *on* the horizon (to a vertical circle passing through the object).

Amplitude (the complement of Azimuth) is the distance from the Prime Vertical, measured *on* the horizon, North or South.

Altitude is the distance from the horizon, measured *on* a vertical circle toward the zenith.

Zenith distance (the complement of Altitude) is the distance from the zenith, measured on a vertical circle, toward the horizon.

The Horizon System is the one commonly used in observations with Mural Circles and Transit Instruments.

II. The Equinoctial System.

(*a*) The Principal Circle is the *Equinoctial.* This is the *Celestial Equator*, or the earth's equator, extended to the Celestial Sphere.

(*b*) Subordinate Circles.—These are the *Hour Circles* (Right Ascension Meridians) and the Declination Parallels. The Hour Circles are thus located. The Equinoctial is divided into 360°, equal to twenty-four hours of motion—thus making 15° equal to one hour of motion. Through these divisions run twenty-four meridians, each constituting an hour of motion (time) or 15° of space. The Hour Circles may be conceived as meridians of terrestrial longitude (15° apart) extended to the Celestial Sphere. (See Colures, p. 40.)

The Declination Parallels are small circles parallel to the Equinoctial; or they may be conceived as the parallels of terrestrial latitude extended to the Celestial Sphere.

(*c*) The POINTS are the *Celestial Poles* and the *Equinoxes.* The *Celestial Poles* are the points where the axis of the earth extended pierces the Celestial Sphere, and are the extremities of the celestial axis, just as the poles of the earth are the extremities of the earth's axis. The North Point is marked very nearly by the North Star, and every direction *from* that is reckoned South, and every direction *toward* that is reckoned North, however it may conflict with our ideas of the points of the compass.

The *Equinoxes* are the points where the Equinoctial and the Ecliptic (the sun's apparent path through the heavens) intersect.

(*d*) The MEASUREMENTS are *Right Ascension* (R. A.), *Declination*, and *Polar Distance.*

Right Ascension is distance from the Vernal Equinox, measured *on* the equinoctial eastward. R. A. corresponds to terrestrial longitude, and may extend to 360° East, instead of 180° as on the earth. R. A. is never measured westward. The starting point is the meridian passing through the vernal equinox; as the meridian passing through Greenwich is the point from which terrestrial longitude is measured.

Declination is distance from the equinoctial, measured *on* any vertical circle or meridian North or South. It corresponds to terrestrial latitude.

Polar distance (the complement of Declination) is

the distance from the Pole, measured on a vertical circle.

The Equinoctial System is largely used by modern astronomers, and accompanies the Equatorial Telescope, Sidereal Clock, and Chronographs of the best Observatories.

III. The Ecliptic System.

(*a*) The Principal Circle is the *Ecliptic.* This is the earth's orbit about the sun, or the apparent path of the sun in the heavens. It is inclined to the equinoctial 23° 28′, which measures the inclination of the Earth's Equator to its orbit, and is called the *obliquity of the ecliptic.*

(*b*) The Subordinate Circles are *Circles of Celestial Longitude*, the *Colures*, and *Parallels of Celestial Latitude.*

The *Circles of Celestial Longitude* are now less employed. They are measured on the Ecliptic, as circles of Right Ascension (R. A.) are now measured on the Equinoctial.

The *Colures* are two principal meridians; the *Equinoctial Colure* is the meridian passing through the equinoxes; the *Solstitial Colure* is the meridian passing through the solstitial points.

The *Parallels of Celestial Latitude* are now little used, but are small circles drawn parallel to the ecliptic, as parallels of declination are now drawn parallel to the equinoctial.

(*c*) The Points are the *Poles of the Ecliptic*, the *Equinoxes*, and the *Solstices*.

The *Poles of the Ecliptic* are the points where the axis of the earth's orbit meets the Celestial Sphere. (Little used.)

The *Equinoxes* are the points where the ecliptic intersects the equinoctial. The place where the sun crosses the equinoctial* in going North, which occurs about the 21st of March, is called the Vernal Equinox. The place where the sun crosses the equinoctial in going South, which occurs about the 21st of September, is called the Autumnal Equinox. The *Solstices* are the two points of the ecliptic most distant from the Equator; or they may be considered to mark the sun's furthest declination, North and South of the equinoctial. The Summer Solstice occurs about the 22d of June; the Winter Solstice occurs about the 22d of December.

(*d*) The Measurements are *celestial longitude* and *latitude*.

Celestial longitude is distance from the Vernal Equinox measured *on* the ecliptic, eastward.

Celestial latitude is distance *from* the ecliptic measured on a Subordinate circle, north or south.

The Zodiac.

A belt of the Celestial Sphere, 8° on each side of the ecliptic, is styled the *Zodiac*. This is of very

* " This is what is commonly called " crossing the line."

high antiquity, having been in use among the ancient Hindoos and Egyptians. The Zodiac is divided into twelve equal parts—of 30° each—called Signs, to each of which a fanciful name is given. The following are the names of the

SIGNS OF THE ZODIAC.

Aries............	♈	Libra............	♎
Taurus	♉	Scorpio..........	♏
Gemini	♊	Sagittarius.......	♐
Cancer	♋	Capricornus......	♑
Leo	♌	Aquarius.........	♒
Virgo	♍	Pisces	♓

"The first, ♈, indicates the horns of the Ram; the second, ♉, the head and horns of the Bull; the barb attached to a sort of letter ♏, designates the Scorpion; the arrow, ♐, sufficiently points to Sagittarius; ♑ is formed from the Greek letters τρ, the two first letters of τράγος, *a goat*. Finally, a balance, the flowing of water, and two fishes, tied by a string, may be imagined in ♎, ♒, and ♓, the signs of Libra, Aquarius, and Pisces."

The Solar System.

" In them hath He set a tabernacle for the sun."

Psalm xix. 4.

THE SOLAR SYSTEM.

THE Solar System is mainly comprised within the limits of the Zodiac. It consists of—

1. The Sun—the centre.
2. The major planets—Vulcan (undetermined), Mercury, Venus, Earth, Mars, Jupiter, Saturn, Uranus, Neptune.
3. The minor planets, at present one hundred and seventeen in number. (The paths of some extend a little outside the Zodiac.)
4. The satellites or moons, eighteen in number, which revolve around the different planets.
5. Meteors and shooting-stars.
6. Nine comets whose orbits have been computed, and over two hundred of which little is known.
7. The Zodiacal Light.

HOW WE ARE TO IMAGINE THE SOLAR SYSTEM TO OURSELVES.—We are to think of it as suspended in space; being held up, not by any visible object, but in accordance with the law of Universal Gravitation discovered by Newton, whereby each planet attracts every other planet and is in turn attracted by all. First, the Sun, a great central globe, so vast as to overcome the attraction of all the planets, and compel them to circle around him; next, the planets, each turning on its axis while it flies around the sun in an elliptical orbit; then, accompanying these, the satellites, each revolving about its own planet, while all whirl in a dizzy waltz about the central orb; next, the comets, rushing across the planetary orbits at irregular intervals of time and space; and finally, shooting-stars and meteors darting hither and thither, interweaving all in apparently inextricable confusion. To make the picture more wonderful still, every member is flying with an inconceivable velocity, and yet with such accuracy that the solar system is the most perfect timepiece known.

THE SUN.

Sign, ⊙, a buckler with its boss.

Distance.—The sun's average distance from the earth is about 91½ million miles. Since the orbit of the earth is elliptical, and the sun is situated at one of its foci, the earth is nearly 3,000,000 miles further from the sun in aphelion than in perihelion. As we attempt to locate the heavenly bodies in space, we are immediately startled by the enormous figures employed. The first number, 91,500,000 miles, is far beyond our grasp. Let us try to comprehend it. If there were air to convey a sound from the sun to the earth, and a noise could be made loud enough to pass that distance, it would require over fourteen years for it to come to us. Suppose a railroad could be built to the sun. An express-train, travelling day and night, at the rate of thirty miles an hour, would require 341 years to reach its destination. Ten generations would be born and would die; the young men would become gray-haired, and their great-grandchildren would forget the story of the beginning of that wonderful journey, and could find it only in history, as we now read of Queen Elizabeth or of Shakspeare; the eleventh generation would see the solar depot at the end of the route. Yet this enormous distance of 91,500,000 miles is used as the unit for expressing celestial distances —as the foot-rule for measuring space; and astronomers speak of so many times the sun's distance as we speak of so many feet or inches.

The Light of the Sun.—This is equal to 5,563 wax-candles held at a distance of one foot from the eye. It would require 800,000 full-moons to produce a day as brilliant as one of cloudless sunshine.

The Heat of the Sun.—The amount of heat we receive annually is sufficient to melt a layer of ice thirty-eight yards in thickness, extending over the whole earth. Yet the sunbeam is only $\frac{1}{300,000}$ part as intense as it is at the surface of the sun. Moreover, the heat and light stream off into space equally in every direction. Of this vast flood only one twenty-three hundred millionth part reaches the earth. It is said that if the heat of the sun were produced by the burning of coal, it would require a layer ten feet in thickness, extending over the whole sun, to feed the flame a single hour. Were the sun a solid body of coal, it would burn up at this rate in forty-six centuries. Sir John Herschel says that if a solid cylinder of ice 45 miles in diameter and 200,000 miles long were plunged, end first, into the sun, it would melt in a second of time.

Apparent size.—It appears to be about a half degree in diameter, so that 360 disks like the sun, laid side by side, would make a half circle of the celestial sphere. It seems a little larger to us in winter than in summer, as we are 3,000,000 miles nearer it. If we represent the luminous surface of the sun when at its average (mean) distance by 1000, the same surface will be represented to us when in aphelion (July) by 940, and when in perihelion (January) by 1072.

Dimensions.—Its diameter is about 850,000 miles.* Let us try to understand this amount by comparison.

A mountain upon the surface of the sun, to bear the same proportion to the globe itself as the Dhawalaghiri of the Himalayas does to the earth, would have to be about six hundred miles high.

Again: Suppose the sun were hollow, and the earth, as in the cut (Fig. 4), placed at the centre, not only would there be room for the moon to revolve in its regular orbit within the shell, but that would stretch off in every direction 200,000 miles beyond.

Its *volume* is 1,245,000 times that of the earth—

* Pythagoras, whose theory of the universe was in so many respects very like the one we receive, believed the sun to be 44,000 miles from the earth, and 75 miles in diameter.

i. e., it would take 1,245,000 earths to make a globe the size of the sun. Its *mass* is 674 times that of all the rest of the solar system. Its *weight* may be expressed in tons thus,

1,910,278,070,000,000,000,000,000,000,

Fig. 4.

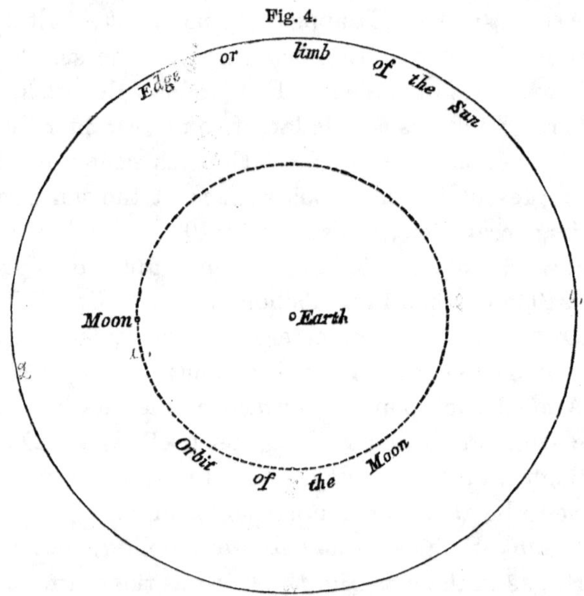

a number which is meaningless to our imagination, but yet represents a force of attraction which holds our own earth and all the planets steadily in their places; while it fills the mind with an indescribable awe as we think of that Being who made the sun, and holds it in the very palm of his hand.

3

The density of the sun is only about one-fourth that of the earth, or 1.43 that of water, so that the weight of a body transferred from the earth to the sun would not be increased in proportion to the comparative size of the two. On account also of the vast size of the sun, its surface is so far from its centre that the attraction is largely diminished, since that decreases, we remember, as the square of the distance. However, a man weighing at the earth's equator 150 lbs., at the sun's equator would weigh about 4,080 lbs.,—a force of attraction that would inevitably and instantly crush him. At the earth's equator a stone falls 16 feet the first second; at the sun's equator it would fall 437 feet.

TELESCOPIC APPEARANCE OF THE SUN: SUN-SPOTS.—We may sometimes examine the sun at early morning or late in the afternoon with the naked eye, and at mid-day by using a smoked glass. The disk will appear to us perfectly distinct and circular, and with no spot to dim its brightness. If we use, however, a telescope of moderate power, taking the precaution to properly shield the eye with a colored eye-piece, we shall find its surface sprinkled with irregular spots, somewhat as shown in the accompanying figure.

Curious opinions concerning solar spots.—The natural purity of the sun seems to have been formerly an article of faith among astronomers, and therefore on no account to be called in question. Scheiner, it is said, having reported to his superior that he had seen spots on the sun's face, was abruptly dismissed with these remarks: "I have read Aristotle's writings from end to end many times, and I assure you I do not find anything in them similar to that which you mention. Go, my son, tranquillize yourself; be assured that what you take for spots are the faults of your glasses or your own eyes."

Fig. 5.

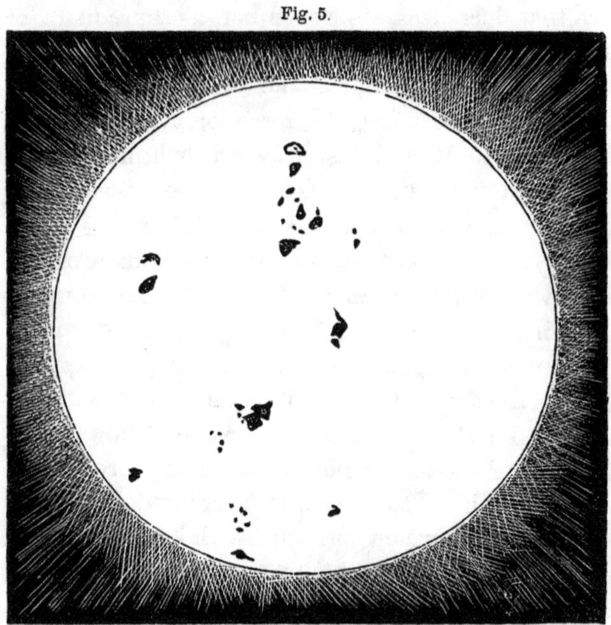

SUN IN TELESCOPE.

Discovery of the solar spots.—They seem to have been noticed as early as 807 A. D., although the telescope was not invented until 1610, and Galileo discovered the solar spots in the following year. We read in the log-book of the good ship Richard of Arundell, on a voyage, in 1590, to the coast of Guinea, that "on the 7, at the going downe of the sunne, we saw a great black spot in the sunne; and the 8 day, both at rising and setting, we saw the like,—which spot to me seeming was about the bignesse of a shilling, being in 5 degrees of latitude, and still there came a great billow out of the souther board."

Number and location of spots.—Sometimes, but rarely, the disk is clear. During a period of ten years, observations were made on 1982 days, on 372 of which there were no spots seen. As many as two hundred spots have been noticed at one time. They are found in two belts, one on each side of the equator, within not less than 8° nor more than 35° of latitude. They seem to herd together—the length of the straggling group being generally parallel to the equator.

The size of the spots.—It is not uncommon to find a spot with a surface larger than that of the earth. Schröter measured one more than 29,000 miles in diameter. Sir J. W. Herschel calculated that one which he saw was 50,000 miles in diameter. In 1843 one was seen which was 14,816 miles across, and was visible to the naked eye for an entire week. On the day of the eclipse in 1858, a spot over 107,000 miles broad was distinctly seen, and attracted general attention in this country. Some who read this paragraph will doubtless recall its appearance. In 1839, Captain Davis saw one which he computed was not less than 186,000 miles long, and had an area of twenty-five billion square miles. If these are deep openings in the luminous atmosphere of the sun, what an abyss must that be at "the bottom of which our earth could lie like a boulder in the crater of a volcano!"

The spots consist of distinct parts.—From the accompanying representation it will be seen that the spots generally consist of one or more dark portions called the *umbra*, and around that a grayish portion styled the *penumbra* (*pene*, almost, and *umbra*, black).—Sometimes, however, umbræ appear without a penumbra, and vice versa. The umbra itself has generally a dense black centre, called the *nucleus*. Besides this, the umbra is sometimes divided by luminous bridges.

Fig. 6.

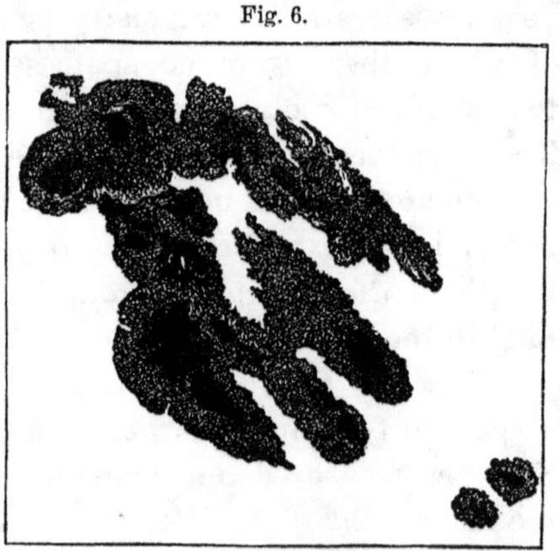

SUN SPOTS.

The spots are in motion.—They change from day to day; but they all have a common movement. About fourteen days are required for a spot to pass across the disk of the sun from the eastern side or *limb* to the western; in fourteen days it reappears, changed in form perhaps, but generally recognizable.

The spots change their rapidity and apparent form as they pass across the disk.—A spot is seen on the eastern limb; day by day it progresses, with a gradually increasing rapidity, until it reaches the centre; it now gradually loses its rapidity, and finally disappears on the western limb. The diagram illustrates the apparent change which takes place in the form. Suppose at first it is of an oval shape; as it approaches the centre it apparently widens and becomes circular. Having passed that point, it becomes more and more oval until it disappears.

Fig. 7.

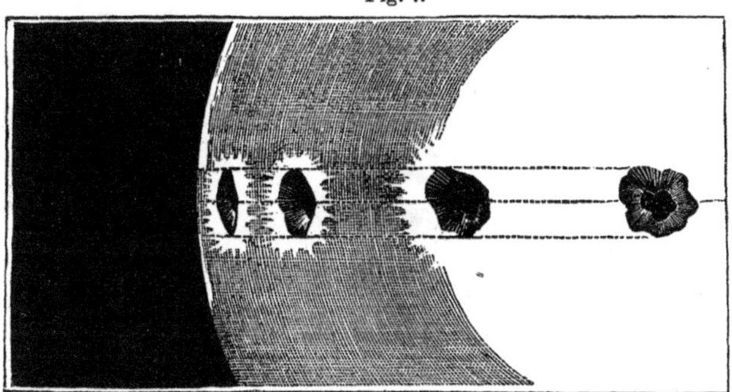

CHANGE IN SPOTS AS THEY CROSS THE DISK.

This change in the spots proves the sun's rotation on its axis.—These changes can be accounted for only on the supposition that the sun revolves on its axis: indeed, they are the precise effects which the laws of perspective demand in that case. About twenty-seven days (27 d., 7 h.) elapse from the appearance of a spot on the eastern limb before it reappears a second time. During this time the earth has gone forward in its orbit, so that the location of the observer is changed; allowing for this, the sun's time of rotation is about twenty-five days (25 d., 8 h., 10 m.: *Langier.*)

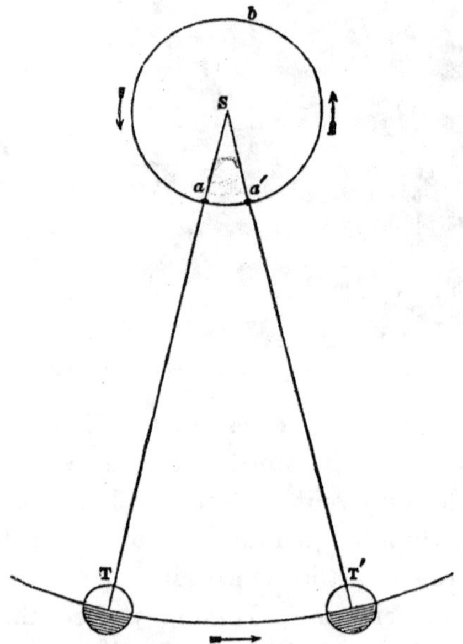

SYNODIC AND SIDEREAL REVOLUTION.

Synodic and sidereal revolution of the spots.—We can easily understand why we make an allowance for the motion of the earth in its orbit. Suppose a solar spot at *a*, on a line passing from the centre of the earth to the centre of the sun. For the spot to pass around the sun and come into that same position again, requires about twenty-seven days. But during this time, the earth has passed on from T to T′. The spot has not only travelled around to *a* again, but also beyond that to *a′*, or the distance from *a* to *a′* more than an entire revolution. To do this requires, as we have said, about two days. A revolution from *a* around to *a′* is called a *synodic*, and one from *a* around to *a* again is called a *sidereal* revolution.

The spots apparently do not always move in straight lines.—Sometimes their path curves toward

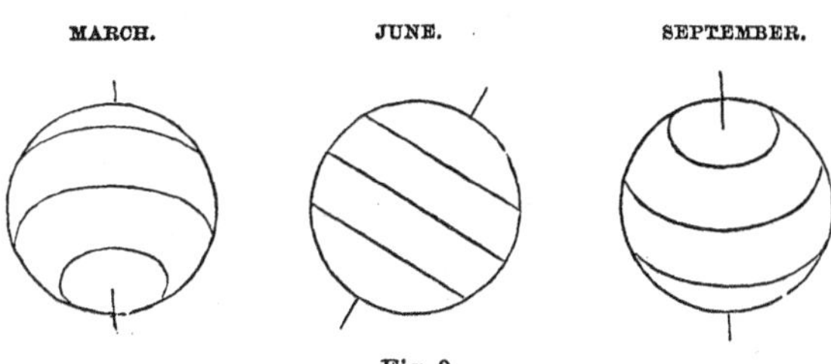

Fig. 9.

the north, and sometimes toward the south, as in the figure. This can be explained only on the supposition that the sun's axis is inclined to the ecliptic (7° 15′).

The spots have a motion of their own.—Besides the motion already named as assigned to the sun's rotation, the spots seem to have a motion of their own, and this fact is undoubtedly the cause of the variation in the estimates made of the time of the sun's revolution on its axis. A spot near the equator

Fig. 10.

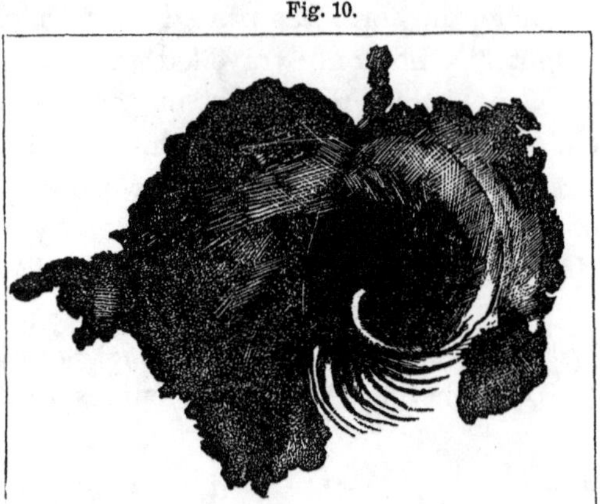

CYCLONE.

performs a synodic revolution in about twenty-five days, while one half way to either pole requires twenty-eight days. One spot was noticed which had a motion three times greater than that of clouds driven along by the most violent hurricane. Again, immense cyclones occasionally pass over the surface with fearful rapidity, producing rotation and sudden changes in the spots. At other times, however, the spots seem "to set sail and move across the disk of the sun like gondolas over a silver sea."

The spots change their real form.—Spots break out and then disappear under the very eye of the astronomer. Wollaston saw one that seemed to be shattered like a fragment of ice when it is thrown on a frozen surface, breaking into pieces, and sliding off in every direction. Sometimes one divides itself into several nuclei, while again several nuclei combine into one. Occasionally a spot will remain for six or eight rotations, while often one will last only half an hour. In one case, Sir. W. Herschel relates that when examining a spot through his telescope, he turned away for a moment, and on looking back it was gone.

The appearance of the spots is periodical.—It is a remarkable fact that the number of spots increases and diminishes through a regular interval of about 11.11 years. These variations seem also to be connected with periodical variations in the aurora, and magnetic earth-currents, which interfere with the telegraph. The regular increase and diminution in the spots was discovered by Schwabe of Prussia, who watched the sun so carefully that it is said, "for thirty years the sun never appeared above the horizon without being confronted by his imperturbable telescope." Besides this, it has now been found that the activity of the sun's spots goes through another regular period of about 56 years. Independently of this conclusion, it has also been discovered that the aurora has a similar period of 56 years.

The spots are influenced by the planets.—They appear to be especially sensitive to the approach of Venus, on account of its nearness, and of Jupiter, because of its size. The area of the spots exposed to view from the earth is uniformly greatest when Venus is on the opposite side of the sun from us, and least when on the same side. When both Venus and Jupiter are on the side of the sun opposite to us, the spots are much larger than when Venus alone is in that position. In part explanation of this influence of the planets, we may suppose that they, in some manner, modify reflection on the disk of the sun exposed to their action, and thus cause a condensation of gases.

The spots do not influence the fruitfulness of the season.—Sir W. Herschel first advanced the idea that years of abundant spots would be years also of plentiful harvest. This is not now generally received. What two years could be more dissimilar than 1859 and 1860? Both abounded in solar spots, yet one was a fruitful year and the other almost one of famine in Europe.

The spots are cooler than the surrounding surface.—It seems that the breaking out of a spot sensibly diminishes the temperature of that portion of the sun's disk. The faculæ, on the other hand, do not increase the temperature. (*Secchi.*)

The spots are depressions below the luminous surface.—This was thought probable before, but is conclusively proved by the photographs of the sun, which have been taken in large numbers of late at Kew Observatory.

Comparative brightness of spots and sun.—If we represent the ordinary brightness of the sun by

1,000, then that of the penumbra would be 469, and that of the nucleus 7. There may be much light and heat radiated by a spot, which seems totally black as compared with the sun : we remember that when we look through even a *Drummond light* at the sun, it appears as a black spot on the disk of that luminary.

Faculæ, willow-leaf, and mottled appearance.—Besides the variety of spots already described, there are other curious appearances worthy of note. Bright ridges or streaks appear, which constitute the most brilliant portions of the sun.—These are called *faculæ*. They vary from barely discernible, softly-gleaming tracts 1,000 miles long, to lofty, piled-up, mountainous regions 40,000 miles long and 4,000 broad. Outside of the spots, the entire disk of the sun is covered with minute shady dots, giving it a mottled appearance not unlike that of the skin of an orange, though less coarse. Under a large telescope the surface seems to be entirely made up of luminous masses, imperfectly separated by dark dots called *pores*. These masses are

Fig. 11.

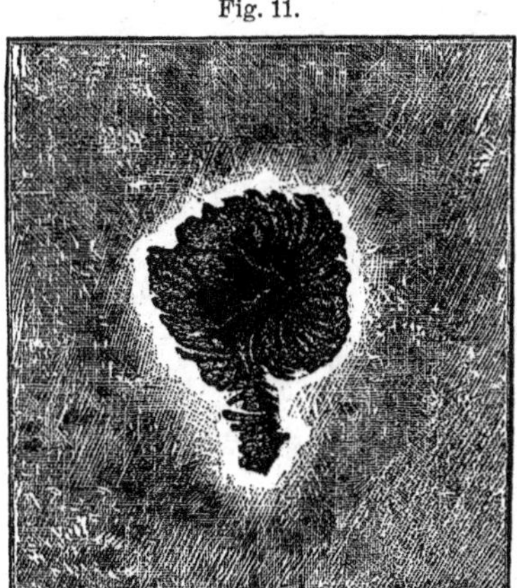

FACULÆ.

said by Mr. Nasmyth to have a "willow-leaf" shape; many observers apply other descriptive terms, such as "rice grains," "untidy circular masses," "things twice as long as broad," "granules," etc. The accompanying cut represents the willow-leafed structure of the luminous surface, and also the "bridges"

Fig. 12.

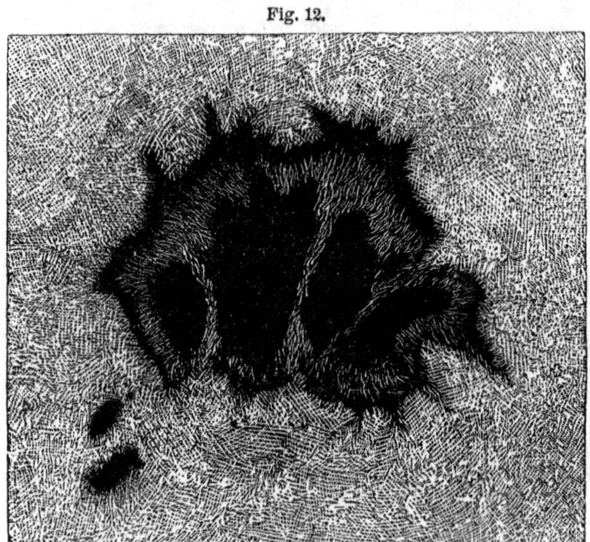

WILLOW-LEAF.

spanning the solar spot. Indeed, it is said that the spots themselves always have their origin in a "*pore*," which appears to slowly increase and assume the blackness of an umbra, after which the penumbra begins to appear.

PHYSICAL CONSTITUTION OF THE SUN.—Of the constitution of the sun, and consequent cause of the solar spots, very little is definitely known. We shall notice the various theories now adopted by different astronomers.

WILSON'S THEORY.—This theory supposes that the sun is composed of a solid, dark globe, surrounded by three atmospheres. The first, nearest the black body of the sun, is a dense, cloudy covering, possessing high reflecting power. The second is called the *photosphere*. It consists of an incandescent gas, and is the seat of the light and heat of the sun. The third, or outer one, is transparent, very like our atmosphere. According to this theory, the spots are to be explained in the following manner. They are simply openings in these atmospheres made by powerful upward currents. At the bottom of these chasms we see the dark sun as a *nucleus* at the centre, and around this the cloudy atmosphere—the *penumbra*. This explains a black spot with its penumbra. Sometimes the opening in the photosphere may be smaller than that in the inner or cloudy atmosphere; in that case there will be a black spot without a penumbra. It will be natural to suppose that when the heated gas of the photosphere or second atmosphere is thus violently rent asunder by an eruption or current from below, luminous ridges will be formed on every side of the opening by the heaped-up gas. This will account for the *faculæ* surrounding the sun-spots. It will be natural, also, to suppose that sometimes the cloudy atmosphere below will close up first over the dark surface of the sun, leaving only an opening through the photosphere, disclosing at the bottom a grayish surface of *penumbra*. We can readily

Fig. 13.

WILSON'S THEORY.

see, also, how, as the sun revolving on its axis brings a spot nearer and nearer to the centre, thus giving us a more direct view of the opening, we can see more and more of the dark body. Then as it passes by the centre the nucleus will disappear, until finally we can see only the side of the fissure, the penumbra, which, in its turn, will pass from our sight. The existence of an outer atmosphere will account for the fact that the sun's margin is not so bright as its centre.

KIRCHHOFF'S THEORY.—This view differs essentially from that of Wilson. It considers the sun as an intensely white-hot solid or fluid body surrounded by a dense atmosphere of flame, filled with substances volatilized by the vivid heat. Changes of temperature take place, which give rise to tornadoes and violent tempests. Descending currents produce openings filled with clouds, which appear as black spots on the sun's disk. A cloud once formed becomes a screen to shield the upper regions from the direct heat of the body of the sun. Thus a lighter cloud is produced, which gives the appearance of a penumbra around the spots.

Spectrum analysis.—The hypothesis just given of the constitution of the sun rests upon the discoveries of the spectroscope. This subject will be treated hereafter under the head of Celestial Chemistry. Wilson's theory is time-honored, but complicated; Kirchhoff's is modern, and partakes of the simplicity of true science.

THE HEAT OF THE SUN.—This subject is not understood. Many theories have been advanced, but none has been generally adopted. Some have supposed the heat is produced by condensation, whereby the size of the sun is being constantly decreased. The dynamic theory accounts for the heat and the solar spots by assuming that there are vast numbers of meteors revolving around the sun, and that these constantly rain down upon the surface of that luminary. Their motion being stopped and changed to heat, feeds this great central fire. Were Mercury to strike the sun in this way, it would generate sufficient heat to compensate the loss by radiation for seven years. Many suppose that the heat of the sun is gradually diminishing. Of this we may be assured, there is enough to support life on our globe for millions of years yet to come.

THE PLANETS.

WE shall describe these in regular order, passing outward from the sun. In this journey we shall examine each planet in turn, noticing its distance, size, length of its year, duration of day and night, temperature of the climate, the number of its moons, and many other interesting facts, showing how much we can know of its world-life in spite of its wonderful distance. We shall encounter the earth in our imaginary wanderings through space, and shall explain many celestial phenomena already partially familiar to us. In all these worlds we shall find traces of the same Divine hand, moulding and directing in conformity to one universal plan. The laws of light and heat will be invariable. The law of gravitation, which causes a stone to fall to the ground, will be found to apply equally to the most distant planets. Even the very elements of which they are composed will be familiar to us, so that a book of natural science published here would, in all its general features, answer for use in a school on Mars or Jupiter.

Characteristics common to the Planets. (*Hind.*) —1. They move in the same invariable direction around the sun; their course, as viewed from the north side of the ecliptic, being contrary to the motion of the hands of a watch.

2. They describe oval or elliptical paths round the sun—not, however, differing greatly from circles.

3. Their orbits are more or less inclined to the ecliptic, and intersect it in two points, which are the nodes—one half of the orbit lying north and the other south of the earth's path.

4. They are opaque bodies like the earth, and shine by reflecting the light they receive from the sun.

5. They revolve upon their axes in the same way as the earth. This we know by telescopic observation to be the case with many planets, and by analogy the rule may be extended to all. Hence they will have the alternation of day and night like the inhabitants of the earth; but their days are of different lengths from our own.

6. Agreeably to the principles of gravitation, their velocity is greatest at those parts of their orbit

which are nearest the sun, and least at the parts which are most distant from it; in other words, they move quickest in perihelion, and slowest in aphelion.

COMPARISON OF THE TWO GROUPS OF THE MAJOR PLANETS. (*Chambers.*)—Separating the major planets into two groups, if we take Mercury, Venus, the Earth, and Mars as belonging to the interior, and Jupiter, Saturn, Uranus, and Neptune to the exterior group, we shall find that they differ in the following respects:

1. The interior planets, with the exception of the earth, are not, so far as we know, attended by any satellite, while the exterior planets *all* have satellites. We can but consider this as one of the many instances to be met with, in the universe, of the beneficence of the Creator, and that the satellites of these remote planets are designed to compensate for the small amount of light their primaries receive from the sun, owing to their great distance from that luminary.

2. The average density of the first group considerably exceeds that of the second, the approximate ratio being 5 : 1.

3. The mean duration of the axial rotations, or mean length of the day of the interior planets, is much longer than that of the exterior; the average in the former case being about twenty-four hours, but in the latter only about ten hours.

THE PROPERTIES OF THE ELLIPSE.—In the figure, S

and S′ are the *foci* of the ellipse; AC is the *major axis;* BD, the *minor* or *conjugate axis;* O, the *centre:* or, astronomically, OA is the *semi-axis-major* or mean distance, OB the *semi-axis-minor:* the ratio of OS to OA is the *eccentricity;* the least distance, SA, is the *perihelion distance;* the greatest distance, SC, the *aphelion distance.*

Fig. 14.

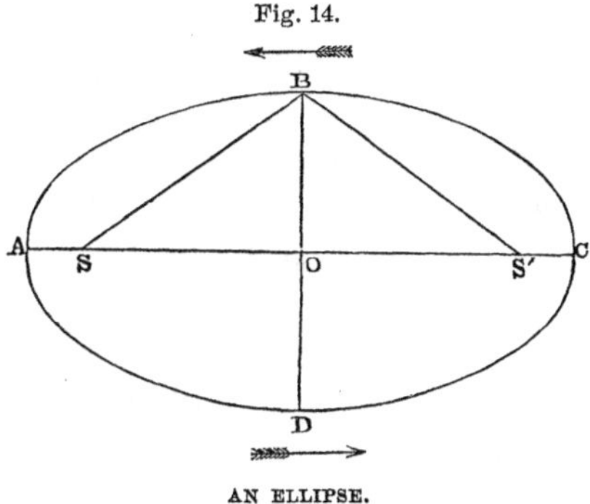

AN ELLIPSE.

CHARACTERISTICS OF PLANETARY ORBIT.—It will not be difficult to follow in the mind the additional

Fig. 15.

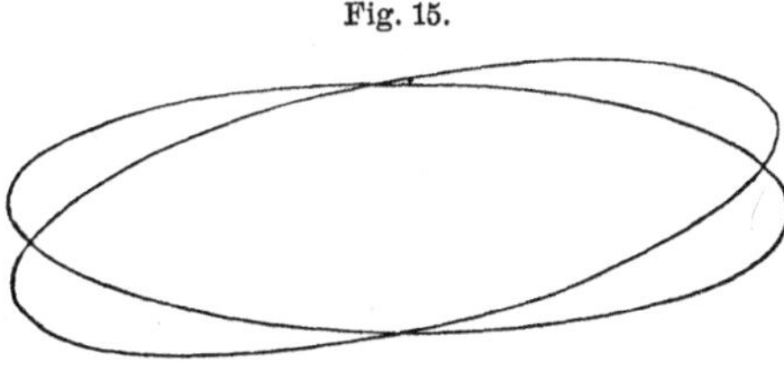

PLANETARY ORBITS.

characteristics of a planet's orbit. The orbit or ellipse just given is laid on a plane surface. Now, incline it slightly, as compared with some other fixed plane ring, as in the cut. The astronomical fixed plane is the ecliptic. Imagine a planet following the inclined ellipse ; at some point it must rise above the level of the fixed plane : this point is called the *ascending node*, and the opposite point of intersection is termed the *descending node.* A line connecting the two nodes is called the *line of the nodes.* The *longitude of the node* is its distance from the first point of Aries, measured on the ecliptic, eastward. In this way we can get a very correct idea of a planetary orbit in space.

COMPARATIVE SIZE OF PLANETS. (*Chambers.*)—The following scheme will assist in obtaining a correct notion of the magnitude of the planetary system. Choose a level field or common ; on it place a globe two feet in diameter for the Sun : Vulcan will then be represented by a small pin's head, at a distance of about 27 feet from the centre of the ideal sun ; Mercury by a mustard-seed, at a distance of 82 feet ; Venus by a pea, at a distance of 142 feet ; the Earth, also, by a pea, at a distance of 215 feet ; Mars by a small pepper-corn, at a distance of 327 feet ; the minor planets by grains of sand, at distances varying from 500 to 600 feet. If space will permit, we may place a moderate-sized orange nearly one-quarter of a mile distant from the starting point to represent Jupiter ; a small orange two-fifths of a mile for Saturn ; a full-sized cherry three-quarters of a mile distant for Uranus ; and lastly, a plum 1¼ miles off for Neptune, the most distant planet yet known. Extending this scheme, we should find that the aphelion distance of Encke's comet would

Fig. 16.

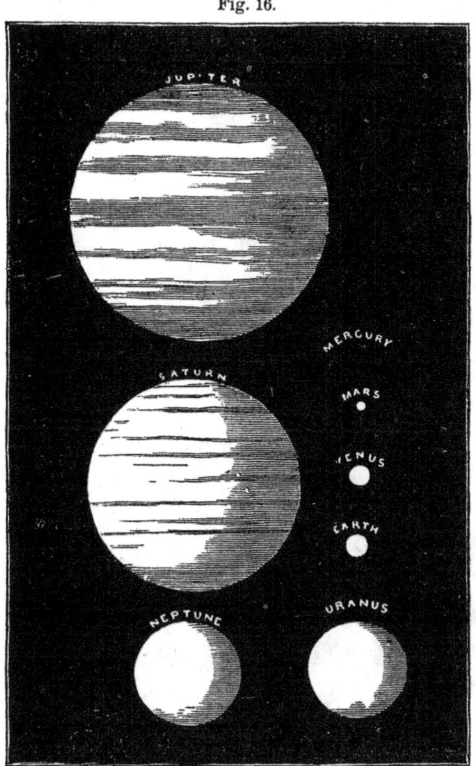

COMPARATIVE SIZE OF PLANETS.

be at 880 feet; the aphelion distance of Donati's comet of 1858 at 6 miles; and the nearest fixed star at 7,500 miles.

According to this scale, the daily motion of Vulcan in its orbit would be $4\frac{2}{3}$ feet; of Mercury, 3 feet; of Venus, 2 feet; of the Earth, $1\frac{7}{8}$ feet; of Mars, $1\frac{1}{2}$ feet; of Jupiter, $10\frac{1}{2}$ inches; of Saturn, $7\frac{1}{2}$ inches; of Uranus, 5 inches; and of Neptune, 4 inches. This illustrates the fact that the orbital velocity of a planet decreases as its distance from the sun increases.

CONJUNCTIONS OF PLANETS.—The grouping together of two or more planets within a limited area of the heavens is a rare event. The earliest record we have is the one of Chinese origin, already mentioned on page 16, wherein it is stated that a conjunction of Mars, Jupiter, Saturn, and Mercury occurred in the

Fig. 17.

VENUS AND JUPITER IN CONJUNCTION, JANUARY 30, 1868.

reign of the Emperor Chuenhio. Astronomers tell us that this actually took place Feb. 28, 2446 B. C., and that they were between 10° and 18° of Pisces. This was before the Deluge, so that the fact must have been afterward calculated and chronicled in their records. In 1859, Venus and Jupiter came so near each other that they appeared to the naked eye as one object. In 1725, Venus, Mercury, Jupiter, and Mars appeared in the same field of the telescope.

ARE THE PLANETS INHABITED?—This question is one which very naturally arises, when we think of the planets as worlds in so many respects similar to our own. We can give no satisfactory answer. Many think that the only object God can possibly have in making any world is to form an abode for man. Our own earth was evidently fitted up, although perhaps not created, for this express purpose. Everywhere about us we find proofs of special forethought and adaptation. Coal and oil in the earth for fuel and light, forests for timber, metals in the mountains for machinery, rivers for navigation, and level plains for corn. Our own bodies, the air, light, and heat are all fitted to each other with exquisite nicety. When we turn to the planets, we do not know but God has other races of intelligent beings who inhabit them, or even entirely different ends to attain. Of this, however, we are assured, that, if inhabited, the conditions on which life is supported vary much from those familiar to us. We shall notice these more especially as we speak of the different planets. We shall see (1) how they differ in light and heat, from seven times our usual temperature to less than $\frac{1}{1000}$; (2) in the intensity of the force of gravity, from $2\frac{1}{2}$ times that of the earth to less than $\frac{1}{20}$; (3) in the constitution of the planet itself, from a density $\frac{1}{4}$ heavier than that of the earth to one nearly that of cork. The temperature sweeps downward through a scale of over

Fig. 18.

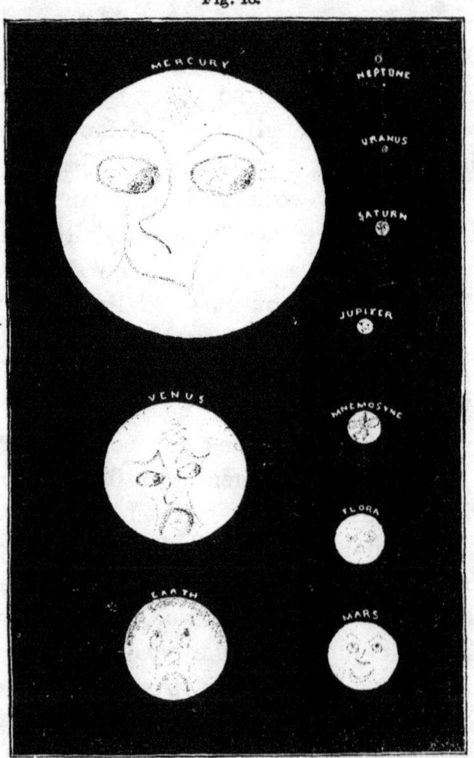

SIZE OF SUN AS SEEN FROM THE PLANETS.

2,000° in passing from Mercury to Uranus. No human being could reside on the former, while we cannot conceive of any polar inhabitant who could endure the intense cold of the latter. At the sun, one of our pounds would weigh 27 pounds; on our moon the pound weight would become only about 2 ounces; while on Vesta, one of the planetoids, a man could easily spring sixty feet in the air and sustain no shock. Yet while we speak of these peculiarities, we do not know what modification oi the atmosphere or physical features may exist even on Mercury to temper the heat, or on Uranus to reduce the cold. With, however, all these diversities, we must admit the power of an all-wise Creator to create beings adapted to the life and the land, however different from our own. The Power that prepared a world for us, could as easily and perfectly prepare one for other races. May it not be that the same love of diversity, which will not make two leaves after the same pattern nor two pebbles of the same size, delights in worlds peopled by races as diverse? While, then, we cannot affirm that the planets are inhabited, analogy would lead us to think that they are, and that the most distant star that shines in the arch of heaven is filled with living beings under the care and government of Him who enlivens the densest forest with the hum of insects, and populates even a drop of water with its teeming millions of animalculæ.

Divisions of the Planets.—The planets are divided into two classes: (1) *Inferior*, or those whose orbits are within that of the earth—viz., Mercury, Venus; (2) *Superior*, or those whose orbits are beyond that of the earth—Mars, Jupiter, Saturn, Uranus, Neptune.

Motions of a Planet as seen from the Sun.—Could we stand at the sun and watch the movements of the planets, they would all be seen to be revolving with different velocities in the order of the zodiacal signs. But to us, standing on one of the planets, itself in motion, the effect is changed. To an observer at the sun all the motions would be real, while to us many are only apparent. The position of a planet, as seen from the centre of the sun, is called its *heliocentric place;* as seen from the centre of the earth, its *geocentric place.* When Venus is at inferior conjunction, an observer at the sun would see it in the opposite part of the heavens from that in which it would appear to him if viewed from the earth.

Motions of an Inferior Planet.—An inferior planet is never seen by us in the part of the sky opposite to the sun at the time of observation. It cannot recede from him as much as 90°, or ¼ the circumference, since it moves in an orbit entirely enclosed by the orbit of the earth. Twice in every revolution it is in conjunction (☌) with the sun,—an *inferior conjunction* (A) when it comes between the earth and the sun, and a *superior conjunction* (B) when the sun lies between it and the earth. (See Fig. 19.)

When the planet attains its greatest distance east or west (as we see it) from the sun, it is said to be at its *greatest elongation*, or in *quadrature* (□).

Fig. 19.

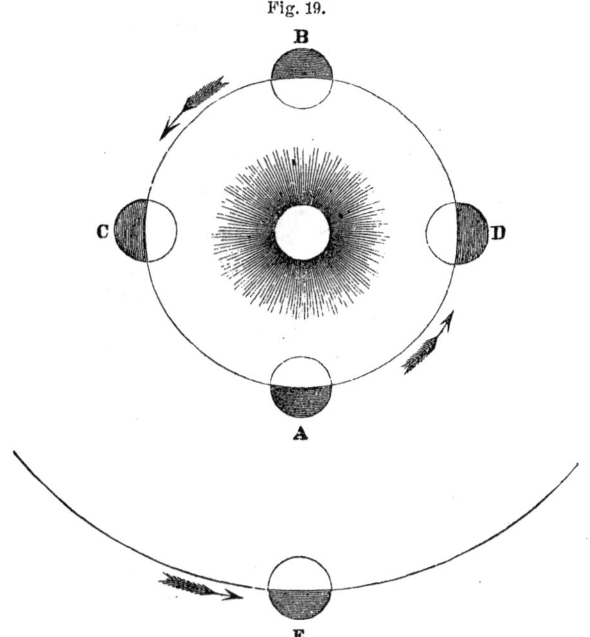

QUADRATURE AND CONJUNCTION.

When passing from B to A it is east of the sun, and from A to B it is west of the sun. When east of the sun, it sets later than the sun, and hence is "evening star:" when west of the sun, it rises earlier than the sun, and hence is "morning star." An inferior planet is never visible when in *superior* conjunction, as its light is then lost in the greater brilliancy of the sun.

When in *inferior* conjunction, it sometimes passes in front of the sun, and appears to us as a round black spot swiftly moving across his disk. This is called a *transit.*

Fig. 20.

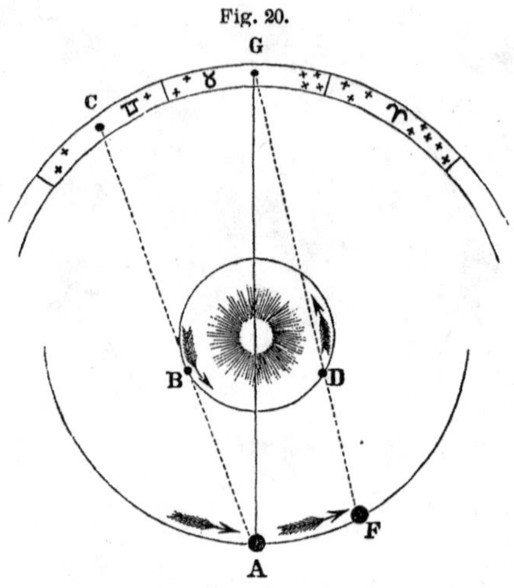

RETROGRADE MOTION.

Retrograde motion of an inferior planet.—Suppose the earth to be at A, and the planet at B. Now, while the earth is passing to F, the planet will pass to D—the arc AF being shorter than BD, because the nearer a planet is to the sun the greater its velocity. While the planet is at B, we locate it a C on the ecliptic, in Gemini; but at D, it appears to us to be at G, in Taurus. So that the planet has retrograded through an entire sign on the ecliptic, while its course all the while has been directly forward in the order of the signs; and to an observer at the sun, such would have been its motion.

Phases of an inferior planet.—An inferior planet presents all the phases of the moon. At superior conjunction, the whole illumined disk is turned toward us; but the planet is lost in the sun's rays: therefore neither Mercury nor Venus ever presents a full circular appearance, like the full moon. A little before or after superior conjunction, an inferior

Fig. 21.

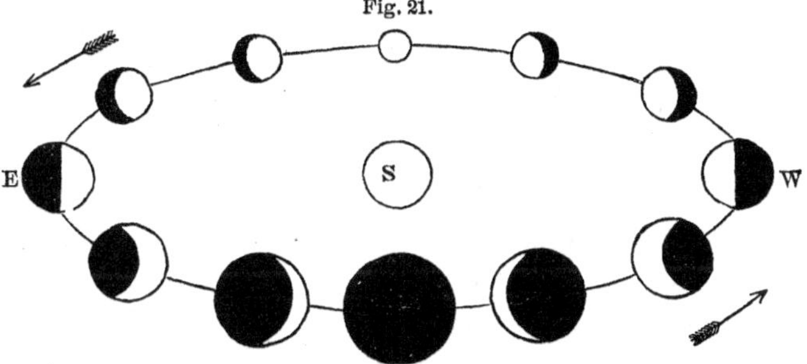

PHASES OF AN INFERIOR PLANET.

planet may be seen with a telescope; but the whole of the light side is not turned toward us, and so the planet appears *gibbous,* like the moon between first quarter and full. In quadrature, the planet shows us only one-half its illumined disk; this decreases, becoming more and more crescent toward inferior conjunction, at which time the unillumined side is toward us.

MOTIONS OF A SUPERIOR PLANET.—The superior planet moves in an orbit which entirely surrounds that of the earth. When the earth is at E (Fig. 22), the planet at L is said to be in *opposition* to the sun. It is then at its greatest distance from him—180°. The planet is on the meridian at midnight while the sun is on the corresponding meridian on the opposite side of the earth; or the planet may be rising when the sun is just setting. When the planet is at N, it is in *conjunction*, and being lost in the sun's rays is invisible to us.

Retrograde motion of a superior planet.—Suppose the earth to be at E and the planet at L, and that we move on to G while the planet passes on to O—the distance EG being longer than LO (just the reverse of what takes place in the movements of the inferior planets); at E, we should locate the planet at P on the ecliptic, in the sign Cancer; but at G, it would appear to us at Q, in the sign Gemini, having apparently retrograded on the ecliptic the distance PQ, while it was all the while moving on in the direct order of the signs. Now, suppose the earth moves on to I and the planet to U, we should then see it at the point W, further on in the ecliptic than Q, which indicates direct motion again, and at some point near Q the planet must have appeared without motion. After this, it will continue direct until the earth has completed a large portion of her orbit, as we shall easily see by imagining various positions of the earth and planet, and then drawing lines as we have just done, noticing whether they indicate direct or retrograde motion. The greater the distance of a planet the less it will retrograde, as we shall perceive by drawing another orbit outside the one represented in the cut, and making the same suppositions concerning it as those we have already explained.

Fig. 22.

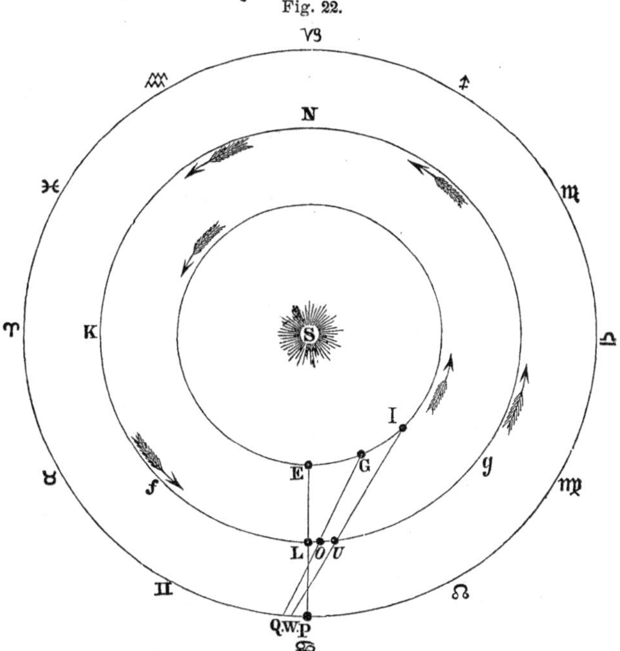

RETROGRADE MOTION OF A SUPERIOR PLANET.

SIDEREAL AND SYNODIC REVOLUTION.—The interval of time required by a planet to perform a revolution from one fixed star back to it again, is termed a *sidereal* revolution (*sidus*, a star).

1. The interval of time between two similar con-

junctions of an inferior planet with the earth and sun is termed a *synodic* revolution. Were the earth at rest, there would be no difference between a sidereal and a synodic revolution, and the planet would come into conjunction twice in each revolution. Since, however, the earth is in motion, it follows that after the planet has completed its sidereal revolution, it must then overtake the earth before they can both come again into the same position with regard to the sun. The faster a planet moves, the sooner it can do this. Mercury, travelling at the greater speed and on an inner orbit, accomplishes it much quicker than Venus. The synodic period always exceeds the sidereal.

2. The interval between two successive conjunctions or oppositions of a superior planet is termed a synodic revolution. Since the earth moves so much faster than any superior planet, it follows that after it has completed a sidereal revolution it must then overtake the planet before they can come again into the same position with regard to the sun. The slower the planet moves, the sooner it can do this. Uranus, making a sidereal revolution in eighty-four years, can be overtaken more quickly than Mars, which makes one in less than two years. It consequently requires over a second revolution to catch up with Mars, $\frac{1}{11}$ of one to overtake Jupiter, and but little over $\frac{1}{100}$ of one to come up with Uranus. Indeed, the earth repasses Neptune in two days after it has finished a sidereal revolution.

PLANETS AS EVENING AND MORNING STARS.—The inferior planets are evening stars from superior to inferior conjunction, and the superior planets from opposition to conjunction. During the other half of their revolutions they are morning stars.

Mercury,	evening	star..........	2	months.
Venus,	"	"	$9\frac{1}{2}$	"
Mars,	"	"	13	"
Jupiter,	"	"	$6\frac{1}{2}$	"
Saturn,	"	"	6	"
Uranus,	"	"	6	"

To avoid filling the text with a multiplicity of figures, many interesting items are condensed in tables at the close of the volume.

VULCAN.

SUPPOSED DISCOVERY.—Le Verrier, having detected an error in the assumed motion of Mercury, suggested, in the fall of 1859, that there may be an interior planet, which is the cause of this disturbance. On this being made public, M. Lescarbault, a French physician, and an amateur astronomer, stated that on March 26 of that year he had seen a dark body pass across the sun's disk, and that this might have been the unknown planet. Le Verrier visited him, and found his instruments rough and home-made, but singularly accurate. His clock was a simple pendulum, consisting of an ivory ball hanging from a nail by a silk thread. His observations were on prescription paper, covered with grease and laudanum. His calculations were chalked on a board, which he planed off to make room for fresh ones. Le Verrier became satisfied that a new planet had been really discovered by this enthusiastic observer, and congratulated him upon his deserved success. On March 20, 1862, Mr. Lummis, of Manchester, England, noticed a rapidly-moving, dark spot, apparently the transit of an inner planet. Many other instances are given of a somewhat similar character. As yet, however, the existence of the planet is not generally conceded. The name Vulcan and the sign of a hammer have been given to it. Its distance from the sun has been estimated at 13,000,000 miles, and its periodic time (its year) at 20 days.

MERCURY.

The fleetest of the gods. Sign, ☿, his wand.

DESCRIPTION.—Mercury is nearest to the sun of any of the definitely known planets. When the sky is very clear, we may sometimes see it, just after the setting of the sun, as a bright sparkling star, near the western horizon. Its elevation increases evening by evening, but never exceeds 30°.* If we watch it closely, we shall find that it again ap-

* This distance varies much, owing to the eccentricity of Mercury's orbit.

proaches the sun and becomes lost in his rays. Some days afterward, just before sunrise, we can see the same star in the east, rising higher each morning, until its greatest elevation equals that which it before attained in the west. Thus the planet appears to slowly but steadily oscillate like a pendulum, to and fro from one side to the other of the sun. The ancients, deceived by this, failed to discover the identity of the two stars, and called the morning star Apollo, the god of day, and the evening star Mercury, the god of thieves, who walk to and fro in the night-time seeking plunder. The Greeks gave to Mercury the additional name of "The Sparkling One." The astrologists looked upon it as the malignant planet. The chemists, because of its extreme swiftness, applied the name to quicksilver. The most ancient account that we have of this planet is given by Ptolemy, in his Almagest; he states its location on the 15th of November, 265 B. C. The Chinese also state that on June 9, 118 A. D., it was near the Beehive, a cluster of stars in Cancer. Astronomers tell us that, according to the best calculations, it was at that date within less than 1° of that group. On account of the nearness of Mercury to the sun, it is difficult to be detected.* It is said that Copernicus, an old man of seventy, lamented in his last moments that, much as he had tried, he had never

* An old English writer by the name of Goad, in 1686, humorously termed this planet, "A squinting lacquey of the sun, who seldom shows his head in these parts, as if he were in debt."

been able to see it. In our latitude and climate, we can generally easily detect it if we watch for it at the time of its greatest elongation or quadrature, as given in the almanac.

MOTION IN SPACE.—It revolves about the sun at a mean distance of 35,000,000 miles. Its orbit is the most eccentric (flattened) of any among the eight principal planets, so that although when in perihelion it approaches to within 28,000,000 miles, in aphelion it speeds away 15,000,000 miles farther, or to the distance of 43,000,000 miles. Being so near the sun, its motion in its orbit is correspondingly rapid—viz., 30 miles per second. At this rate of speed, we could cross the Atlantic Ocean in two minutes. The Mercurial year comprises only about 88 days, or nearly three of our months. Mercury revolves upon its axis in about the same time as the earth, so that the length of the Mercurial day is nearly the same as that of the terrestrial one. Though Mercury thus completes a sidereal revolution around the sun in 88 days, yet to pass from one inferior or superior conjunction to the same again (a synodic revolution) requires 116 days. The reason of this is, as already explained, that when Mercury comes around to the same spot in its orbit again, the earth has gone forward, and it requires 28 days for the planet to overtake us.

DISTANCE FROM THE EARTH.—This varies still more than its distance from the sun. At inferior conjunction it is between the earth and the sun, and its distance from us is the *difference* between the distance of the earth and the planet from the sun: at superior conjunction it is the *sum* of these distances. Its apparent diameter in these different positions varies in the same proportion as the distances, or as three to one. The greatest and least distances vary according as either planet may happen to be in aphelion or perihelion. If at inferior conjunction Mercury is in aphelion and the earth in perihelion, its distance from us is only 90,000,000 – 43,000,000 = 47,000,000 miles. If at superior conjunction Mercury is in aphelion and the earth in aphelion also, its distance from us is 93,000,000 + 43,000,000 = 136,000,000 miles.

DIMENSIONS.—Mercury is about 3,000 miles in *diameter*. Its *volume* is about $\frac{1}{20}$ that of the earth—*i. e.*, it would require twenty globes as large as Mercury to make one the size of the earth, or 25,000,000 to equal the sun. Yet as it is $\frac{1}{4}$ denser than the earth, its weight is nearly $\frac{1}{16}$ that of the earth, and a stone let drop upon its surface would fall $7\frac{1}{2}$ feet the first second. Its specific gravity is about that of tin. A pound weight removed to Mercury would weigh only about seven ounces.

SEASONS.—As Mercury's axis is much inclined from a perpendicular, its seasons are peculiar. There are no distinct frigid zones; but large regions near the poles have six weeks continuous day and torrid heat, alternating with a night of equal length and arctic cold. The sun shines perpendicularly upon the torrid zone only at the equinoxes, while he sinks far toward the southern horizon at one solstice, and as far toward the northern horizon at the other. The equatorial regions, therefore, modify their temperature during each rev-

Fig. 23

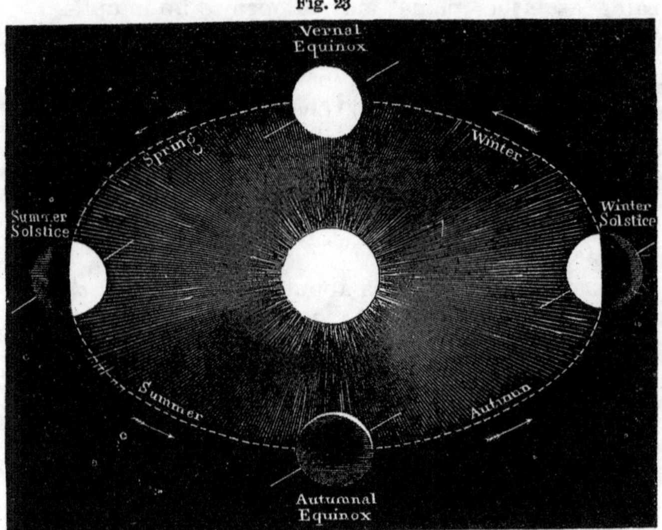

ORBIT AND SEASONS OF MERCURY.

olution from torrid to temperate, and the tropical heat is experienced alternately toward the north and south of what we call the temperate zones. There is no marked distinction of zones as with us, but each zone changes its character twice during the Mercurial year, or eight times during the terrestrial one. An inhabitant of Mercury must be accustomed to the most sudden and violent vicissitudes of temperature. At one time the sun not only thus pours down its vertical rays, and in a few weeks after sinks far down toward the horizon, but, on account of Mercury's elliptical orbit, when in perihelion the planet approaches so near the sun that the heat and light are ten times as great as that we receive, while in aphelion it recedes so as to reduce the amount to four and a half times (the average, however, is seven times),—a temperature sufficient to turn water to steam, and even to melt many of the metals. This entire round of transitions is swept through four times during one terrestrial year. The relative length of the days and nights is much more variable than with us. The sun, apparently seven times as large as it seems to us, must be a magnificent spectacle, and illumine every object with insufferable brilliancy. The evening sky is, however, lighted by no moon.

TELESCOPIC FEATURES.—Under the telescope, Mercury presents all the phases of the moon, from a slender crescent to gibbous, when its light is lost in that of the sun. These phases prove that Mercury is spherical, and shines by the light reflected from the sun. When in quadrature, it can sometimes be detected with a telescope in daylight. Being an inferior planet, we can never see it when full, and hence the brightest, nor when nearest the earth, as then its dark side is turned toward us. Owing to the dazzling light, and the vapors almost always hanging around our horizon, this planet has not received much attention of late; the cuts here given, and the remarks concerning its physical features, are based upon the observations of the older astronomers. It is thought by some to have a dense *atmosphere* loaded with clouds, which would materially diminish the intensity of the sun, and perhaps make Mercury quite habitable. Sir W. Herschel, however, emphatically denies this, and asserts that the atmosphere is too insignificant to be detected. There are some dark bands about its equator. It has lofty mountains, which intercept the light of the sun, and deep valleys plunged in shade. One mountain has been ascertained to be about ten miles in height, which is $\frac{1}{300}$ of the diameter of the planet. The height of the Dhawalaghiri of the Himalayas is less than 29,000 feet, or $\frac{1}{1400}$ part of the earth's diameter.

VENUS.

The Queen of Beauty. Sign ♀, a looking-glass.

Description.—Venus, the next in order to Mercury, is the most brilliant of all the planets. When visible before sunrise, she was called by the ancients Phosphorus, Lucifer, or the Morning Star, and when she shone in the evening after sunset, Hesperus, Vesper, or the Evening Star. She presents the same appearances as Mercury. Owing, however, to the greater diameter of her orbit, her apparent oscillations are nearly 48° east and west of the sun,* or about 18° more than those of Mercury. She is therefore seen much earlier in the morning and much later at night. She is "morning star" from inferior to superior conjunction, and "evening star" from superior to inferior conjunction. She is the most brilliant about five weeks before and after inferior conjunction, at which time the planet is bright enough to cast a shadow at night. If, in addition, at this time of greatest brilliancy, Venus is at or near her highest north latitude, she may be seen with the naked eye in full daylight.† This occurs once in eight years, in which interval the earth and planet return to the same situation in their orbits; eight complete revolutions of the earth about the sun occupying nearly the same time as thirteen of Venus. This happened last in February, 1862. A less degree of brilliancy is attained once in twenty-nine months, under somewhat the same circumstances.

MOTION IN SPACE.—Unlike Mercury, Venus has an orbit the most circular of any of the principal

* This distance varies but little, owing to the slight eccentricity of Venus's orbit.

† Arago relates that Bonaparte, upon repairing to the Luxembourg, when the Directory was about to give him a *fête*, was much surprised at seeing the multitude paying more attention to the heavens above the palace than to him or his brilliant staff. Upon inquiry, he learned that these curious persons were observing with astonishment a star which they supposed to be that of the Conqueror of Italy. The emperor himself was not indifferent when his piercing eye caught the clear lustre of Venus smiling upon him at midday.

planets. Her mean distance from the sun is about 66,000,000 miles, which varies at aphelion and perihelion within the limits of a half million miles against 15,000,000 miles in the case of the former planet. She makes a complete revolution around the sun in about 225 days, at the mean rate of 22 miles per second; hence her year is equal to about seven and one half of our months. This is a *sidereal* revolution, as it would appear to an observer at the sun, but a *synodic* revolution is 584 days. Mercury, we remember, catches up with the earth in 28 days after it reaches the point where it left the earth at the last inferior conjunction. But it takes Venus nearly two and a half revolutions to overtake the earth and come into the same conjunction again. This grows out of the fact that Venus has a longer orbit to travel through, and moves only about one-fifth faster than the earth, while Mercury travels nearly twice as fast. The planet revolves upon its axis in about 24 hours; so the day does not differ in length essentially from ours.

DISTANCE FROM THE EARTH.—The distance of Venus from the earth, like that of Mercury, when in inferior conjunction, is the difference between the distances* of these two planets from the sun, and when in superior conjunction the sum of these distances.

---

* Let the pupil calculate the distances of the earth and Venus from each other, when in perihelion and aphelion, as in the case of Mercury, (See tables in Appendix.)

The figure represents its apparent dimensions at the extreme, mean, and least distances from us. The variation is nearly as the numbers 10, 18, and 65. It would be natural to think that the planet is the brightest when the nearest, and thus the largest,

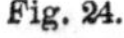

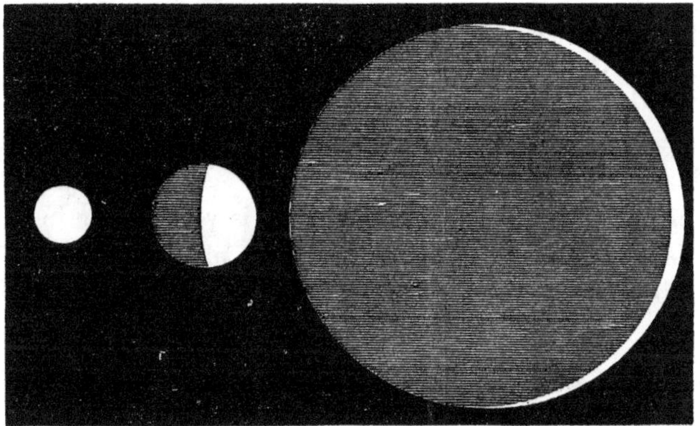

EXTREME, MEAN, AND LEAST APPARENT SIZE OF VENUS.

but we should remember that then the bright side is toward the sun, and the unillumined side toward us. Indeed, at the period of greatest brilliancy of which we have spoken, only about one-fourth of the light is visible. At this time, however, many observers have noticed the entire contour of the planet to be of a dull gray hue, as seen in the cut.

DIMENSIONS.—Venus is about 7,500 miles in *diameter*. The *volume* of the planet is about four-fifths that of the earth, while the *density* is about the same. A stone let fall upon its surface would fall 14 feet in the first second: a pound weight removed to its equator would weigh about five-sixths of a pound. From this we see that the force of gravity does not decrease exactly in proportion to the size of the planet, any more than it increases with the mass of the sun. The reason of this is, that the body is brought nearer the mass of the small planet, and so feels its attraction more fully than when far out upon the extreme circumference of a large body,—the attraction increasing as the square of the distance from the particles decreases.

SEASONS.—As the axis of Venus is very much inclined from a perpendicular, its seasons are similar to those of Mercury. The torrid and temperate

Fig. 25

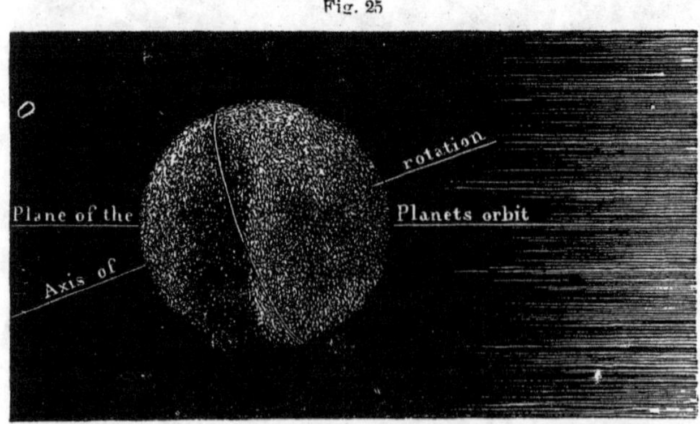

VENUS AT ITS SOLSTICE.

zones overlap each other; the polar regions having alternately at one solstice a torrid temperature, and at the other a prolonged arctic cold. The inequality of the nights is very marked. The heat and light are double that of the earth, while the circular form of its orbit gives nearly an equal length to its four seasons.

If the inclination of its axis is 75°, as some astronomers hold, its tropics must be 75° from the equator, and its polar circles 75° from the poles. The torrid zone is, therefore, 150° in width. The torrid and frigid zones interlap through a space of 60°, midway between the equator and poles.

TELESCOPIC FEATURES.—Venus, being an interior planet, presents, like Mercury, all the phases of the moon. This fact was discovered by Galileo, and was among the first achievements of his telescopic observations. It had been argued against the Copernican system that, if true, Venus should wax and wane like the moon. Indeed, Copernicus himself boldly declared that if means of seeing the planets more distinctly were ever invented, Venus would be found to present such phases. Galileo, with his telescope, proved this fact, and, by overthrowing that objection, again vindicated the Copernican theory. This planet is not sensibly *flattened* at the poles. It is thought to have a dense, cloudy atmosphere. This was established by the fact that at the transit of Venus over the sun in 1761 and 1769, a faint ring of light was observed to surround the black disk of the planet. The evidence of an atmosphere, as well as of mountains, rests very much upon the peculiar appearance attending its crescent shape.

(1.) The luminous part does not end abruptly; on the contrary, its light diminishes gradually, which diminution may be entirely explained by the twilight on the planet. The existence of an atmosphere

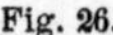

CRESCENT AND SPOTS OF VENUS.

which diffuses the rays of light into regions where the sun has already set, has hence been inferred. Thus, on Venus, the evenings, like ours, are lighted by twilight, and the mornings by dawn. (2.) The edge of the enlightened portion of the planet is uneven and irregular. This appearance is doubtless the effect of shadows cast by mountains. *Spots* have been noticed on its disk which are considered to be traceable to clouds. Indeed, Herschel thinks that we never see the real body of the planet, but only its atmosphere loaded with vapors, which may mitigate the glare of the intense sunshine.

SATELLITES.—Venus is not known to have any moon.

THE EARTH.

Sign, ⊕, a circle with Equator and Meridian.

The Earth is the next planet we meet in passing outward from the sun. To the beginner, it seems strange enough to class our world among the heavenly bodies. *They* are brilliant, while *it* is dark and opaque; they appear light and airy, while it is solid and firm; we see in it no motion, while they are constantly changing their position; they seem mere points in the sky, while it is vast and extended. Yet at the very beginning we are to consider the earth as a planet shining brightly in the heavens, and appearing to other worlds as a star does to us: we are to learn that it is in motion, flying through its orbit with inconceivable velocity; that it is not fixed, but hanging in space, held by an invisible power of gravitation which it cannot evade; that it is small and insignificant beside the mighty globes that so gently shine upon us in the far-off sky; that our earth is only one atom in a universe of worlds, all firm and solid, and equally well fitted to be the abode of life.

DIMENSIONS.—The earth is not "round like a ball," but flattened at the poles. Its form is that of an oblate spheroid. Its polar diameter is about 7,899 miles, and its equatorial about 7,925½. The compression is, therefore, about 26½ miles. (See table in Appendix.) If we represent the earth by a globe one yard in diameter, the polar diameter would be one-tenth of an inch too long. It has been recently

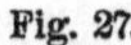

Fig. 27.

THE EARTH IN SPACE.

shown that the equator itself is not a perfect circle, but is somewhat flattened, since the diameter which pierces the meridian 14° east of Greenwich is two miles longer than the one at right angles to it. The circumference of the earth is about 25,000 miles. Its density is about $5\frac{1}{2}$ times that of water. Its weight is

6,069,000,000,000,000,000,000 tons.

The inequalities of its surface, arising from buildings, valleys, mountains, etc., have been likened to the roughness on the rind of an orange. This is not an exaggeration. On a globe sixteen inches in diameter, the land, to be in proportion, should be represented by the thinnest writing-paper, the hills by small grains of sand, and elevated ranges by thick drawing-paper. To represent the deepest wells or mines, a scratch might be made that would be invisible except with a glass. The water in the ocean could be shown by a brush dipped in color and lightly drawn over the bed of the sea.

THE ROTUNDITY OF THE EARTH.—This is shown in various ways, among which are the following: (1) By the fact that vessels have sailed around the earth;*

---

* It is curious, in connection with this well-known fact, to recall the arguments urged by the Spanish philosophers against the reasoning of Columbus, when he assured them that he could arrive at Asia just as certainly by sailing west as east. "How," they asked, "can the earth be round? If it were, then on the opposite side the rain would fall upward, trees would grow with their branches down, and everything would be topsy-turvy. Every object on its surface would certainly fall off; and if a ship by sailing west should get around

(2) when a ship is coming into port we see the masts first; (3) the shadow of the earth on the moon is circular; (4) the polar star seems higher in the heavens as we pass north; (5) the horizon expands as we ascend an eminence.* If we climb to the top of a hill, we can see further than when on the plain at its foot. Our eyesight is not improved; it is only because ordinarily the curvature of the earth shuts off the view of distant objects, but when we ascend to a higher point, we can see farther over the side of the earth. The curvature is eight inches per mile, $2^2 \times 8^{\text{in.}} = 32$ inches for two miles, $3^2 \times 8^{\text{in.}}$ for three miles, etc. An object of these respective heights would be just hidden at these distances.

APPARENT AND REAL MOTION.—In endeavoring to understand the various appearances of the heavenly bodies, it is well to remember how in daily life we transfer motion. On the cars, when in rapid movement, the fences and trees seem to glide by us,

---

there, it would never be able to climb up the side of the earth and get back again. How can a ship sail up hill?"

* The history of aëronautic adventure affords a curious illustration of this same principle. The late Mr. Sadler, the celebrated aëronaut, ascended on one occasion in a balloon from Dublin, and was wafted across the Irish Channel, when, on his approach to the Welsh coast, the balloon descended nearly to the surface of the sea. By this time the sun was set, and the shades of evening began to close in. He threw out nearly all his ballast, and suddenly sprang upward to a great height, and by so doing brought his horizon to *dip* below the sun, producing the whole phenomenon of a western sunrise. Subsequently descending in Wales, he, of course, witnessed a second sunset on the same evening.

while we sit still and watch them pass. On a bridge, when we are at rest, we follow the undulations of the waves, until at last we come to think that they are stationary and we are sweeping down stream. "In the cabin of a large vessel going smoothly before the wind on still water, or drawn along a canal, not the smallest indication acquaints us with the 'way it is making.' We read, sit, walk, as if we were on land. If we throw a ball into the air, it falls back into our hand; if we drop it, it alights at our feet. Insects buzz around us as in the free air, and smoke ascends in the same manner as it would do in an apartment on shore. If, indeed, we come on deck, the case is in some respects different; the air, not being carried along with us, drifts away smoke and other light bodies such as feathers cast upon it, apparently in the opposite direction to that of the ship's progress; but in reality *they* remain at rest, and we leave them behind in the air."*

Diurnal Revolution of the Earth around its Axis.—The earth, in constantly turning from west

* "And what is the earth itself but the good ship we are sailing in through the universe, bound round the sun; and as we sit here in one of the 'berths,' we are unconscious of there being any 'way' at all upon the vessel. On deck, too, out in the open air, it's all the same as long as we keep our eyes on the ship; but immediately we look over the sides—and the horizon is but the 'gunwale' of our vessel—we see the blue tide of the great ocean around us go drifting by the ship, and sparkling with its million stars as the waters of the sea itself sparkle at night between the tropics."

to east, elevates our horizon above the stars on the west, and depresses it below the stars on the east. As the horizon appears to us to be stationary, we assign the motion to the stars, thinking those on the west which it passes over and hides to have sunk below it or *set*, and imagining those on the east it has dropped below to have moved above it or *risen*. So, also, the horizon is depressed below the sun, and we call it *sunrise;* it is elevated above the sun, and we call it *sunset*. We thus see that the diurnal movement of the sun by day and stars by night is a mere optical delusion—that here as elsewhere we simply transfer motion. This seems easy enough for us to understand, because the explanation makes it so simple; but it was the "stone of stumbling" to ancient astronomers for two thousand years. Copernicus himself, it is said, first thought of the true solution while riding on a vessel and noticing how he insensibly transferred the movement of the ship to the objects on the shore. How much grander the beautiful simplicity of this theory than the cumbersome complexity of the old Ptolemaic belief!

Diurnal motion of the Sun.—The explanation just given illustrates the apparent motion of the sun, and the cause of day and night. Suppose S to be the sun. E, the earth, turning upon its axis EF from west to east, has half its surface only illuminated at one time by the sun. To a person at D, the sun is in the horizon and day commences, the luminary appearing to rise higher and higher in the heavens with a westerly motion, as the observer is carried forward easterly by the earth's diurnal rotation to A, where he has the sun in his

Fig. 28.

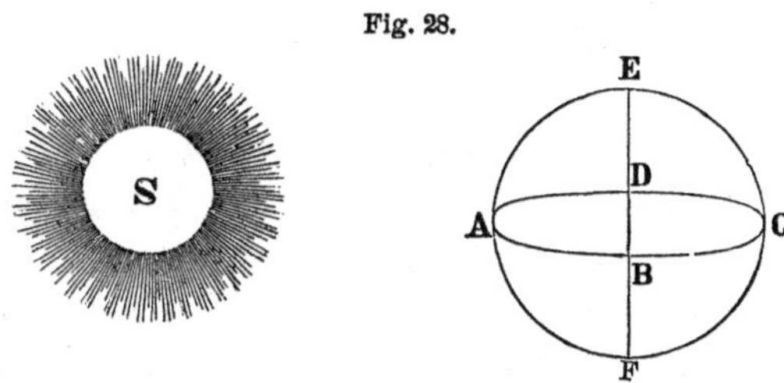

DAILY MOTION OF THE SUN.

meridian, and it is consequently noon. The sun then begins to decline in the sky until the spectator arrives at B, where it sets, or is again in the horizon on the west side, and night begins. He moves on to C, which marks his position at midnight, the sun being then on the meridian of places on the opposite part of the earth, and he is then brought round again to D, the point of sunrise, when another day commences. (Hind.)

The unequal rate of diurnal motion.—Different points upon the surface of the earth revolve with different velocities. At the two poles the speed of rotation is nothing, while at the equator it is greatest, or over 1,000 miles per hour. At Quito, the circle of latitude is much longer than one at the mouth of the St. Lawrence, and the velocities vary in the same proportion. The former place moves at the rate of about 1,038 miles per hour; the latter, 450 miles. In our latitude (41°) the speed is about 780 miles per hour. We do not perceive this wonderful velocity with which we are flying through the air, because the air moves with us.* Yet were the earth suddenly to stop its rotation, the terrible shock would, without doubt, destroy the entire race of man, and we, with houses, trees, rocks, and even the oceans, in one confused mass, would be hurled headlong into space. On the other hand, were the rate of rotation to increase, the length of the day would be proportionately shortened, and the weight of all bodies decreased by the centrifugal force thus produced. Indeed, if the rotary movement should become swift enough to

---

* An ingenious inventor once suggested that we should utilize the earth's rotation, as the most simple and economical, as well as rapid mode of locomotion that could be conceived. This was to be accomplished by rising in a balloon to a height inaccessible to aerial currents. The balloon, remaining immovable in this calm region, would simply await the moment when the earth, rotating underneath, would present the place of destination to the eyes of travellers, who would then descend. A well-regulated watch and an exact knowledge of longitudes would thus render travelling possible from east to west, all voyages north or south naturally being interdicted. This suggestion has only one fault; it supposes that the atmospheric strata do not revolve with the earth. Upon that hypothesis, since we rotate in our latitude with the velocity of 333 yards in a second, there would result a wind in the contrary direction ten times more violent than the most terrible hurricane. Is not the absence of such a state of things a convincing proof of the participation of the atmospheric envelope in the general movement? (Guillemin.)

reduce the day to 84 minutes, or about $\frac{1}{17}$ its present length, the force of gravity would be overcome, and, at the equator, all bodies would be without weight; if the speed were still further increased, loose bodies would fly off from the earth like water from a grindstone when swiftly turned, while we should be compelled to constantly "hold on" to avoid sharing the same fate. But against such a catastrophe we are assured by the immutability of God's laws. "He is the same yesterday, to-day, and forever." The earth has not varied in its revolution $\frac{1}{100}$ of a second in 2,000 years.

Fig. 29.

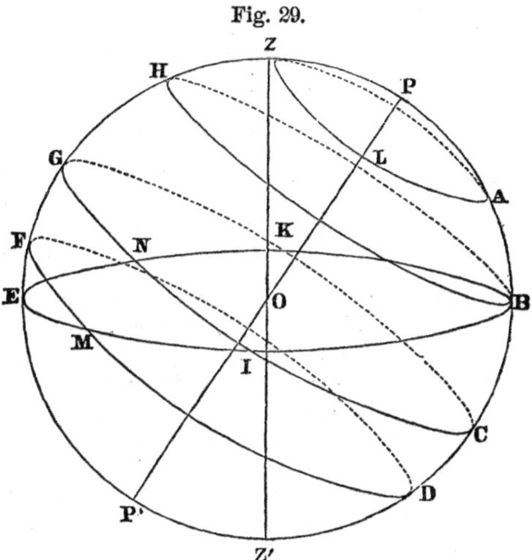

Unequal diurnal orbits of the stars.—Let O represent our position on the earth's surface, E Z B our meridian; E I B K our horizon; P and P′ the north and south poles, Z the zenith, Z′ the nadir; and GICK the celestial equator. Now PB, it will be seen, is the elevation of the north pole above the horizon, or the latitude of the place. Suppose we should see a star at A, on the meridian below the pole. The earth revolves in the direction GIC; the star will therefore move along A L to Z, when it is on the meridian above the pole. It continues its course along the dotted line around to A again, when it is on the meridian below the pole, having made a complete circuit around the pole, but not having descended below our horizon. A star rising at B would just touch the horizon; one at I would move on the celestial equator, and would be above the horizon as long time as it is below—twelve hours in each case; a star rising at M, would just come above the horizon and set again at N.

Unequal diurnal velocities of the stars.—The stars appear to us to be set in a concave shell which rotates daily about the earth. As different parts of the earth *really* revolve with varying velocities, so the stars *appear* to revolve at different rates of speed. Those near the pole, having a small orbit, revolve very slowly, while those near the celestial equator move at the greatest speed.

Appearance of the stars at different places on the earth.—Were we placed at the north pole, Polaris would be directly overhead, and the stars would seem to pass around us in circles parallel to the horizon, and increasing in diameter from the upper to lower ones. Were we placed at the equator, the pole-star would be at the horizon, and the stars would move in circles exactly perpendicular to the horizon, and decreasing in diameter, north and south from those in the zenith, while we could see one half of the path of each star. Were we placed in the southern hemisphere, the circumpolar stars would rotate about the south pole, and the others in circles resembling those in our sky, only the points of direction would be reversed to correspond with the pole. Were we placed at the south pole, the appearance would be the same as at the north pole, except that there is no star to mark the direction of the earth's axis.

Motion of the Earth in Space about the Sun.—The earth revolves in an elliptical path about the sun at a mean distance of 91½ million of miles. This path is called the *ecliptic;* its eccentricity, which is greater than that of the orbit of Venus, changes about $\frac{4}{100,000}$ per century, so that in time the orbit would become circular, were it not that after the lapse of some thousands of years, the eccentricity will begin to increase again, and will thus vary through all ages within definite, although yet undetermined limits. Its entire circumference is nearly 600,000,000 miles, and the earth pursues this wonderful journey at the rate of 18 miles per second. This revolution of the earth about the sun gives rise to various phenomena, of which we shall now proceed to speak.

1. *Change in the appearance of the heavens in different months.*—This is the natural result of the revolution of the earth about the sun. In Fig. 30, suppose

Fig. 30.

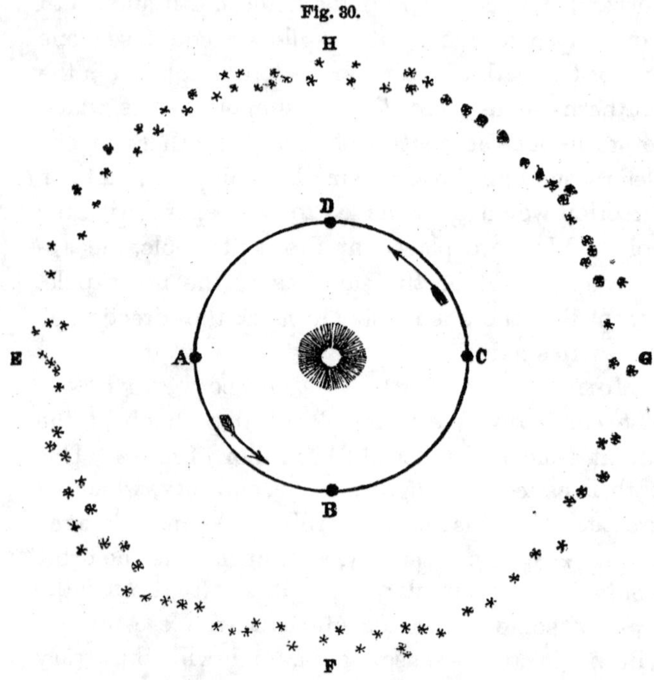

APPEARANCE OF THE HEAVENS IN DIFFERENT SEASONS.

A B C D to be the orbit of the earth, and E F G H the sphere of the fixed stars, surrounding the sun in every direction. When our globe is at A, the stars about E are on the meridian at midnight. Being seen from the earth in the opposite quarter to the sun, they are most favorably placed for observation. The stars at G, on the contrary, will be invisible, for the sun intervenes between them and the earth: they are on the meridian of the spectator about the same time as the sun, and are always hidden in his rays. In three months the earth has passed over one-fourth of her orbit, and has arrived at B. Stars about F now appear on the meridian at midnight. while those at E, which previously occupied their places, have descended toward the west and are becoming lost in the sun's refulgence, while those about G are just coming into sight in the east. In three months more the earth is situated at C, and stars about G shine in the midnight sky, those at F having, in their turn, vanished in the west. Stars at E are on the meridian at noon, and consequently hidden in daylight; and those about H are just escaping from the sun's rays, and commencing their appearance in the east. One revolution of the earth brings the same stars again on the meridian at midnight. Thus it is that the earth's motion round the sun as a centre explains the varied aspect of the heavens in the summer and winter skies. (Hind.)

2. *Yearly path of the sun through the heavens.*—We have spoken of the diurnal motion of the sun. We now speak of its *second* apparent motion—its yearly path among the stars.* If we look at the accom-

* This yearly movement of the sun among the fixed stars is not as apparent to us as his daily motion, because his superior

panying plate (Fig. 31), we can see how the motion of the earth in its orbit is also transferred to the sun, and causes him to appear to us to travel in a fixed path through the heavens. When the earth is in any part of the ecliptic, the sun seems to us to be in the point directly opposite. For example, when the earth is in Libra (♎)*—autumnal equinox—the sun is in Aries (♈)—vernal equinox; when the sun enters the next sign, Taurus (♉), the earth in fact has passed on to Scorpio (♏). Thus as the earth moves through her orbit, the sun seems to pass through the same path along the opposite side of the ecliptic, making the entire circuit of the heavens in the year, and returning at the end of that time to the same place among the stars. If the earth could leave a shining line as it passes through its orbit about the sun, we should see the sun apparently moving along this same line on the opposite side of the circle. We therefore define the *ecliptic* as the *real orbit of the earth about the sun*, or the *apparent path of the sun through the heavens*. The ecliptic crosses the celestial equator at two points. These are called the *equinoxes*.

---

light blots out the stars. But if we notice a star at the western horizon just at sunset, we can tell what constellation the sun is then in: now wait two or three nights, and we shall find that star is set, and another has taken its place. Thus we can trace the sun through the year in his path among the fixed stars.

* When we say "the earth is in Libra," we mean that a spectator placed at the sun would see the earth in that part of the heavens which is occupied by the sign Libra.

3. *An apparent movement of the sun, north and south.*—Having now spoken of the apparent *diurnal* and *annual* motions of the sun, there yet remains a *third* motion, which has doubtless oftentimes attracted our attention. In summer, at midday, the sun is high in the heavens; in the winter, quite low, near the southern horizon. In summer he is a long time above the horizon; in the winter, a short time. In summer he rises and sets north of the east and west points; in winter, south of the east and west points. This subject is so intimately connected with the next, that we shall understand it best when taken in connection with it.

4. CHANGE OF THE SEASONS.—VARIATION IN LENGTH OF DAY AND NIGHT.—By closely studying the accompanying illustration and imagining the various positions of the earth in its orbit, let us try to understand the several points.

I. *Obliquity of the ecliptic.*—The axis of the earth is inclined $23\frac{1}{2}°$ from a perpendicular to its orbit. This angle is called the obliquity of the ecliptic.

II. *Parallelism of the axis.*—In all parts of the orbit, the axis of the earth is parallel to itself and constantly points toward the North Star.* This is only an instance of what is very familiar to us all. Nature reveals to us nothing more permanent than the axis of rotation in anything that is rapidly turned. It is its rotation which keeps a boy's hoop

* There is a slight variation from this, which we shall soon notice.

Fig. 31.

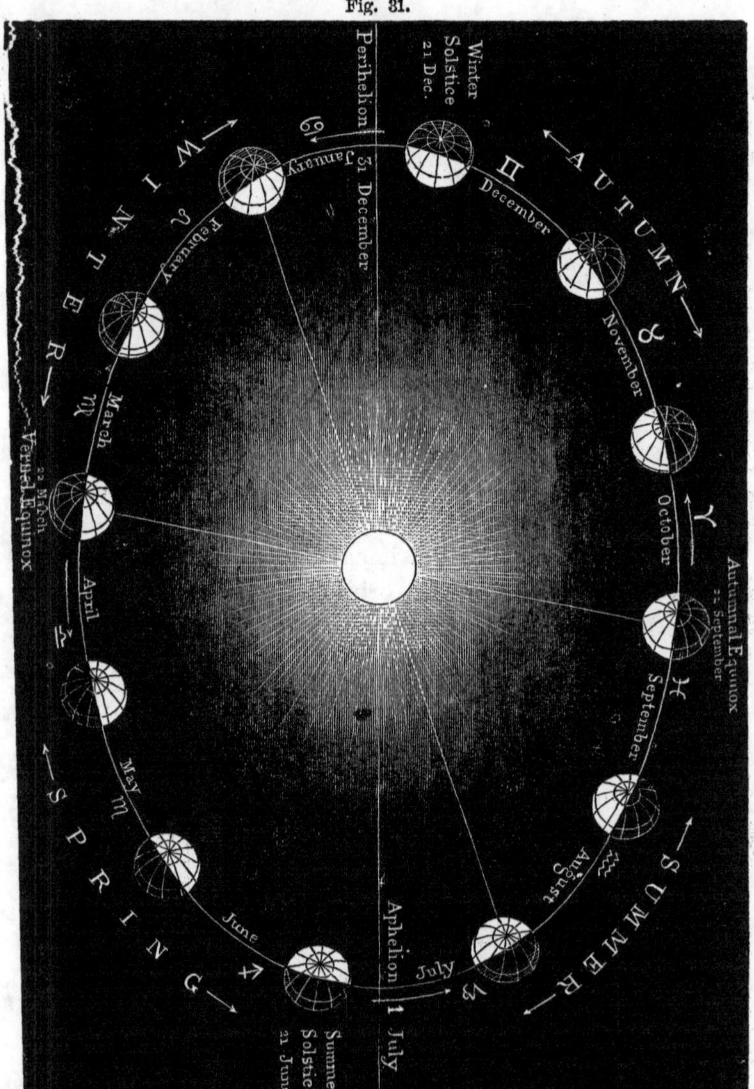

THE ORBIT OF THE EARTH.

from falling. For the same reason a quoit retains its direction when whirled, and it will keep in the same plane at whatever angle it may be thrown. A man slating a roof wishes to throw a slate to the ground; he simply whirls it, and as it descends it will strike on the edge without breaking. As long as a top spins there is no danger of its falling, since its tendency to preserve parallel its axis of rotation is greater than the attraction of the earth. This wonderful law would lead us to think that the axis of the earth always points in the same direction, even if we did not know it from direct observation.

III. *The rays of the sun strike the various portions of the earth, when in any position, at different angles.*—Example. When the earth is in Libra, and also when in Aries, the rays strike vertically at the equator, and more and more obliquely in the northern and southern hemispheres, as the distance from the equator increases, until at the poles they strike almost horizontally. This variation in the direction of the rays produces a corresponding variation in the intensity of the sun's heat and light at different places, and accounts for the difference between the torrid and polar regions.

IV. *As the earth changes its position the angle at which the rays strike any portion is varied.*—Example. Take the earth as it enters Capricornus (♑) and the sun in Cancer (♋) He is now overhead, 23½° *north* of the equator. His rays strike less obliquely in the northern hemisphere than when the earth was in Libra. Let six months elapse: the earth is now in Cancer and the sun in Capricornus; and he is overhead, $23\frac{1}{2}°$ *south* of the equator. His rays strike less obliquely in the southern hemisphere than before, but in the northern hemisphere more obliquely. These six months have changed the direction of the sun's rays on every part of the earth's surface. This accounts for the difference in temperature between summer and winter.

V. *The Equinoxes.*—At the equinoxes one half of each hemisphere is illuminated: hence the name Equinox (*æquus*, equal, and *nox*, night). At these points of the orbit the days and nights are equal over the entire earth,* each being twelve hours in length.

VI. *Northern and southern hemispheres unequally illuminated.*—While one half of the earth is constantly illuminated, at all points in the orbit except the equinoxes the proportion of the northern or southern hemisphere which is in daylight or darkness varies. When more than half of a hemisphere is in the light, its days are longer than the nights, and *vice versa.*

VII. *The seasons and the comparative length of days and nights in the South Temperate Zone, at any specified time, are the reverse of those in the North Temperate Zone, except at the Equinoxes, where the days and nights are of equal length.*

---

* Except a small space at each pole.

VIII. *The earth at the Summer Solstice.*—When the earth is at the summer solstice, about the 21st of June, the sun is overhead 23½° north of the equator, and if its vertical rays could leave a golden line on the surface of the earth as it revolves, they would mark the Tropic of Cancer. The sun is at its furthest northern declination, ascends the highest it is ever seen above our horizon, and rises and sets 23½° north of the east and west points. It seems now to stand still in its northern and southern course, and hence the name Solstice (*sol*, the sun, *sto*, to stand). The days in the north temperate zone are longer than the nights. It is our summer, and the 21st of June is the longest day of the year. In the south temperate zone it is winter, and the shortest day of the year. The circle that separates day from night extends 23½° beyond the north pole, and if the sun's rays could in like manner leave a golden line on that day, they would trace on the earth the Arctic Circle. It is the noon of the long six months polar day. The reverse is true at the Antarctic Circle, and it is there the midnight of the long six months polar night.

IX. *The earth at the Autumnal Equinox.*—The earth crosses the aphelion point the 1st of July, when it is at its furthest distance from the sun, which is then said to be in *apogee*. The sun each day rising and setting a trifle further toward the south, passes through a lower circuit in the heavens. We reach the autumnal equinox the 22d of September. The sun being now on the equinoctial, if its vertical rays could leave a line of golden light, they would mark on the earth the circle of the equator. It is autumn in the north temperate zone and spring in the south temperate zone. The days and nights are equal over the whole earth, the sun rising at 6 A. M. and setting at 6 P. M., exactly in the east and west where the equinoctial intersects the horizon.

X. *The earth at the Winter Solstice.*—The sun after passing the equinoctial—"crossing the line," as it is called—sinks lower toward the southern horizon each day. We reach the winter solstice the 21st of December. The sun is now directly overhead 23½° south of the equator, and if its rays could leave a line of golden light they would mark on the earth's surface the Tropic of Capricorn. It is at its furthest southern declination, and rises and sets 23½° south of the east and west points. It is our winter, and the 21st of December is the shortest day of the year. In the south temperate zone it is summer, and the longest day of the year. The circle that separates day from night extends 23½° beyond the south pole, and if the sun's rays in like manner could leave a line of golden light they would mark the Antarctic Circle. It is there the noon of the long six months polar day. At the Arctic Circle the reverse is true; the rays fall 23½° short of the north pole, and it is there the midnight of the long six months polar night. Here again the sun appears to us to stand still a day or two before retracing its course, and it is therefore called the Winter Solstice.

XI. *The earth at the Vernal Equinox.*—The earth reaches its *perihelion* about the 31st of December. It is then nearest the sun, which is therefore said to be in *perigee.* The sun rises and sets each day further and further north, and climbs up higher in the heavens at midday. Our days gradually increase in length, and our nights shorten in the same proportion. On the 21st of March* the sun reaches the equinoctial, at the vernal equinox. He is overhead at the equator, and the days and nights are again equal. It is our spring, but in the south temperate zone it is autumn.

XII. *The yearly path finished.*—The earth moves on in its orbit through the spring and summer months. The sun continues its northerly course, ascending each day higher in the heavens, and its rays becoming less and less oblique. On the 21st of June it again reaches its furthest northern declination, and the earth is at the summer solstice. We have thus traced the yearly path, and noticed the course of the changing seasons, with the length of the days and nights. The same series has been repeated through all the ages of the past, and will be till time shall be no more.

XIII. *Distance of the earth from the sun varies.*—

* The precise time of the equinoxes and solstices varies each year, but within a small limit.

We notice, from what we have just seen, that we are nearer the sun by 3,000,000 miles in winter than in summer. The obliqueness with which the rays strike the north temperate zone at that time prevents our receiving any special benefit from this favorable position of the earth.

XIV. *Southern summer.*—The inhabitants of the south temperate zone have their summer while the earth is in perihelion, and the sun's rays are about $\frac{1}{30}$ warmer than when in aphelion, our summer-time. This will perhaps partly account for the extreme heat of their season. Herschel tells us that he has found the temperature of the surface soil of South Africa 159° F. Captain Sturt, in speaking of the extreme heat of Australia, says that matches accidentally dropped on the ground were immediately ignited. The southern winters, for a similar reason, are colder; and this makes the average *yearly temperature* about the same as ours.

XV. *Extremes of heat and cold not at the solstices.*—We notice that we do not have our greatest heat at the time of the summer solstice, nor our greatest cold at the winter solstice. After the 21st of June, the earth, already warmed by the genial spring days, continues to receive more heat from the sun by day than it radiates by night: thus its temperature still increases. On the other hand, after the 21st of December the earth continues to become colder, because it loses more heat during the night than it receives during the day.

XVI. *Summer longer than winter.*—As the sun is not in the centre of the earth's orbit, but at one of its foci, that portion of the orbit which the earth passes through in going from the vernal to the autumnal equinox comprises more than one-half the entire ecliptic. On this account the summer is longer than the winter. The difference is still further enhanced by the variation in the earth's velocity at aphelion and perihelion. The annexed table gives the mean duration of the seasons :

Seasons.	Days.	Seasons.	Days.
Spring..........	92.9	Autumn	89.7
Summer.........	93.6	Winter	89.0

The difference of time in the earth's stay in the two portions of the ecliptic, as will be seen from the above, is 7.8 days.

XVII. *Varying velocity of the earth.*—We can see, by looking at the plate, that the velocity of the earth must vary in different portions of its orbit. When passing from the vernal equinox to aphelion, the attraction of the sun tends to check its speed ; from that point to the autumnal equinox, the attraction is partly in the direction of its motion, and so increases its velocity. The same principle applies when going to and from perihelion.

XVIII. *Curious appearance of the sun at the north pole.*—"To a person standing at the north pole, the sun appears to sweep horizontally around the sky every twenty-four hours, without any perceptible variation in its distance from the horizon. It is, however, slowly rising, until, on the 21st of June, it is twenty-three degrees and twenty-eight minutes above the horizon, a little more than one-fourth of the distance to the zenith. This is the highest point it ever reaches. From this altitude it slowly descends, its track being represented by a spiral or screw with a very fine thread; and in the course of three months it worms its way down to the horizon, which it reaches on the 22d of September. On this day it slowly sweeps around the sky, with its face half hidden below the icy sea. It still continues to descend, and after it has entirely disappeared it is still so near the horizon that it carries a bright twilight around the heavens in its daily circuit. As the sun sinks lower and lower, this twilight grows gradually fainter, till it fades away. December 21st, the sun is 23° 28′ below the horizon, and this is the midnight of the dark polar winter. From this date the sun begins to ascend, and after a time it is heralded by a faint dawn, which circles slowly around the horizon, completing its circuit every twenty-four hours. This dawn grows gradually brighter, and on the 22d of March the peaks of ice are gilded with the first level rays of the six months day. The bringer of this long day continues to wind his spiral way upward, till he reaches his highest place on the 21st of June, and his annual course is completed."

XIX. *Results, if the axis of the earth were perpendicular to the ecliptic.*—The sun would then always appear to move through the equinoctial. He would rise and set every day at the same points on the horizon, and pass through the same circle in the heavens, while the days and nights would be equal the year round. There would be near the equator a fierce torrid heat, while north and south the climate would melt away into temperate spring, and, lastly, into the rigors of a perpetual winter.

XX. *Results, if the equator of the earth were perpendicular to the ecliptic.*—Were this the case, to a spectator at the equator, as the earth leaves the vernal equinox, the sun would each day pass through a smaller circle, until at the summer solstice he would reach the north pole, when he would halt for a time and then slowly return in an inverse manner.

In our own latitude, the sun would make his diurnal revolutions in the way we have just described, his rays shining past the north pole further and further, until we were included in the region of perpetual day, when he would seem to wind in a spiral course up to the north pole, and then return in a descending curve to the equator.

PRECESSION OF THE EQUINOXES.—We have spoken of the equinoxes as if they were stationary in the orbit of the earth. Over two thousand years ago, Hipparchus found that they were falling back along the ecliptic. Modern astronomers fix the rate at about 50″ of space annually. If we mark either point in the ecliptic at which the days and nights are equal over the earth, which is where the plane of the earth's equator passes exactly through the centre of the sun, we shall find the earth the next year comes back to that position 50″ (20 m. 20 s. of time) earlier. This remarkable effect is called the *Precession of the Equinoxes*, because the position of the equinoxes in any year precedes that which they occupied the year before. Since the circle of the ecliptic is divided into 360°, it follows that the time occupied by the equinoctial points in making a complete revolution at the rate of 50.2″ per year is 25,816 years.

Results of the Precession of the Equinoxes.—In Fig. 31, we see that the line of the equinoxes is not at right angles to the ecliptic. In order that the plane of the terrestrial equator should pass through the sun's centre 50″ earlier, it is necessary that the plane itself should slightly change its place. The axis of the earth is always perpendicular to this plane, hence it follows that the axis is not rigorously parallel to itself. It varies in direction, so that the north pole describes a small circle in the starry vault twice 23° 28′ in diameter. To illustrate this, in the cut we suppose that after a series of years the position of the earth's equator has changed from *efh* to *g* K *l*. The inclination of the axis of the earth, C P, to C Q, the pole of the ecliptic, remains unchanged; but as it must turn with the equator, its position is moved from C P to C P′, and it passes slowly around through a portion of a circle whose centre is C Q. The direction of this motion is the same as that of the hands of a watch, or just the reverse of that of the revolution of the earth itself. The position of the north pole in the heavens is therefore gradually but almost insensibly changing. It is now distant from the north polar star about $1\frac{1}{2}°$. It will continue to approach

Fig. 32.

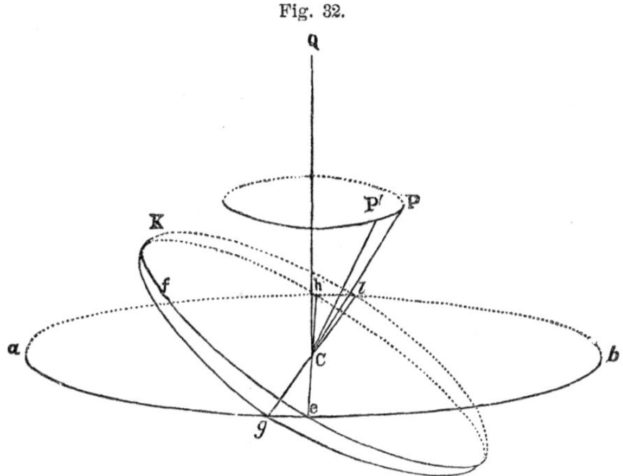

CHANGE OF EARTH'S EQUATOR AND AXIS.

it until they are not more than half a degree apart. In 12,000 years Lyra will be our polar star: 4,500 years ago the polar star was the bright star in the constellation Draco. As the right ascension of the stars is reckoned eastward from the vernal equinox along the equinoctial, the precession of the equinoxes increases the R. A. of the stars 50″ per year. On this account, star maps must be accompanied by the date of their calculations, that they may be corrected to correspond with this annual variation. The constellations are fixed in the heavens, while the signs of the zodiac are not; they are simply abstract divisions of the ecliptic which move with the equinox. When named, the sun was in both the sign and constellation Aries, at the time of the vernal equinox; but since then the equinoxes have retrograded nearly a whole sign, so that now while the sun is in the sign Aries on the ecliptic, it corresponds to the constellation Pisces in the heavens. Pisces is therefore the first constellation in the zodiac. (See Fig. 72.)

Causes of the Precession of the Equinoxes.—Before commencing the explanation of this phenomenon, it is necessary to impress upon the mind a few facts. 1. The earth is not a perfect sphere, but is swollen at the equator. It is like a perfect sphere covered with padding, which increases constantly in thickness from the poles to the equator; this gives it a turnip-like shape. 2. The attraction of the sun is

Fig. 33.

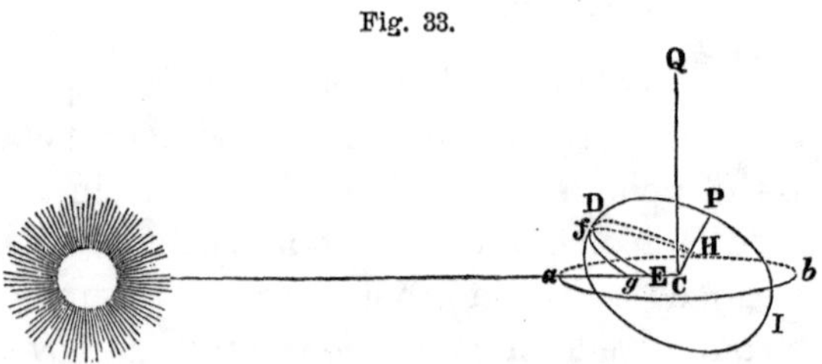

INFLUENCE OF THE SUN ON A MOUNTAIN NEAR THE EQUATOR.

greater the nearer a body is to it. 3. The attraction is not for the earth as a mass, but for each particle separately. In the figure, the position of the earth at the time of the winter solstice is represented. P is the north pole, *a b* the ecliptic, C the centre of the earth, C Q a line perpendicular to the ecliptic, so that the angle Q C P equals the obliquity of the ecliptic. In this position the equatorial padding we have spoken of—the ring of matter about the equator—is turned, not exactly toward the sun, but is elevated above it. Now the attraction of the sun pulls the part D more strongly than the centre; the tendency of this is to bring D down to *a*. In the same way the attraction for C is greater than for I, so it tends to draw C away from I, and as at the same time D tends toward *a*, to pull I up toward *b*. The tendency of this, one would think, would be to change the inclination of the axis C P toward C Q, and make it more nearly perpendicular to the ecliptic. This would be the result if the earth were not revolving upon its axis. Let us consider the case of a mountain near the equator. This, if the sun did not act upon it, would pass through the curve H D E in the course of a semi-revolution of the earth. It is nearer the sun than the centre C is; the attraction therefore tends to pull the mountain downward and tilt the earth over, as we have just described; so the mountain will pass through the curve H *f g*, and instead of crossing the ecliptic at E it will cross at *g* a little sooner than it otherwise would. The same influence, though in a less degree, obtains on the opposite side of the earth. The mountain passes around the earth in a curve nearer to *b*, and crosses the ecliptic a little earlier. The same reasoning will apply to each mountain and to all the protuberant mass near the equatorial regions. The final effect is to turn slightly the earth's equator so that it intersects the ecliptic sooner than it would were it not for this attraction. At the summer solstice the same tilting motion is produced. At the equinoxes the earth's equator passes directly through the centre of the sun, and therefore there is no tendency to change of position. As the axis C P must move with the equator, it slowly revolves, keeping its inclination unchanged, around C Q, the pole of the ecliptic, describing, in about 26,000 years, a small circle twice 23° 28′ in diameter.

Precession illustrated in the spinning of a top.—This motion of the earth's axis is most singularly illustrated in the spinning of a top, and the more remarkably because there the forces are of an opposite character to those which act on the earth, and so produce an opposite effect. We have seen that if the earth had no rotation, the sun's attraction on the "padding" at the equator would bring C P nearer to C Q, but that in consequence of this rotation the effect really produced is that C P, the earth's axis,

Fig. 34.

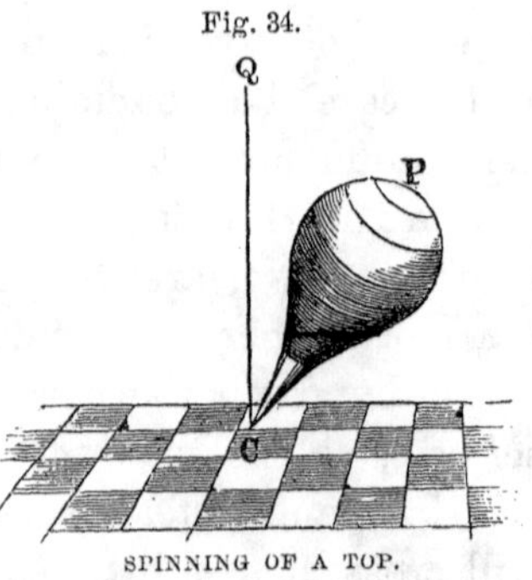

SPINNING OF A TOP.

slowly *revolves around* C Q, the pole of the heavens, in a direction *opposite* to that of rotation.

In Fig. 34, let C P be the axis of a spinning top, and C Q the vertical line. The direct tendency of the earth's attraction is to bring C P *further from* C Q (or to make the top fall), and if the top were not spinning this would be the result; but in consequence of the rotary motion the inclination does not sensibly alter (until the spinning is retarded by friction), and so C P slowly revolves around C Q in the *same direction* as that of rotation.

NUTATION (*nutatio*, a nodding).—We have noticed the sun as producing precession; the moon has, however, treble its influence; for although her mass is not $\frac{1}{20,000,000}$ part that of the sun, yet she is 400 times nearer and her effect correspondingly greater. (See p. 168.) The moon's orbit does not lie parallel to the ecliptic, but is inclined to it. Now the sun attracts the moon, and disturbs it as he would the path of the mountain we have just supposed, and the effect is the same—viz., the intersections of the moon's orbit with the ecliptic travel backward, completing a revolution in about 18 years. During half of this time the moon's orbit is inclined to the ecliptic in the same way as the earth's equator; during the other half it is inclined in the opposite way. In the former state, the moon's attractive tendency to tilt the earth is very small, and the precession is slow; in the latter, the tendency is great, and precession goes on rapidly.

The consequence of this is, that the pole of the earth is irregularly shifted, so that it travels in a slightly curved line, giving it a kind of "wabbling" or "nodding" motion, as shown—though greatly exaggerated—in Fig. 35. The obliquity of the ecliptic, which we consider 23° 28′, is the *mean* of the irregularly curved line and is represented by the dotted circle.

Fig. 35.

PATH OF THE NORTH POLE IN THE HEAVENS.

Periodical change in the obliquity of the ecliptic.—Although it is sufficiently near for all general purposes to consider the obliquity of the ecliptic invariable, yet this is not strictly the case. It is subject to a small but appreciable variation of about 46″ per century. This is caused by a slow change of the position of the earth's orbit, due to the attraction of the planets. The effect of this movement is to gradually diminish the inclination of the earth's equator to the ecliptic (the obliquity of the ecliptic). This will continue for a time, when the angle will as gradually increase; the extreme limit of change being only 1° 21′. The orbit of the earth thus vibrates backward and forward, each oscillation requiring a period of 10,000 years. This change is so intimately blended, in its effect upon the obliquity of the ecliptic, with that caused by precession and nutation, that they are only separable in theory; in point of fact, they all combine to produce the waving motion we have already described. As a consequence of this variation in the obliquity of the ecliptic, the sun does not come as far north nor decline as far south as at the Creation, while the position of all the terrestrial circles—Tropic of Cancer, Capricorn, Arctic, etc.—is constantly but slowly changing. Besides this, it tends to vary slightly the comparative length of the days and nights, and, as the obliquity is now diminishing, to equalize them. As the result of this variation in the position of the orbit, some stars which were formerly just south of the ecliptic are now north of it, and others that were just north are now a little further north; thus the latitude of these stars is gradually changing.

Change in the major axis (line of apsides) of the earth's orbit.—Besides all the changes in the position of the earth in its orbit due to precession, the line connecting the aphelion and perihelion points of the orbit itself is slowly moving. The consequence of this is a variation in the length of the seasons at different periods of time. In the year 4089 B. C., about the supposed epoch of the creation, the earth was in perihelion at the autumnal equinox, so that the summer and autumn seasons were of equal length, but shorter than the winter and spring seasons, which were also equal.* In the

* There is much discrepancy in the views held concerning the Great Year of the astronomers, as it is often called. (See 14 Weeks in Geology, pp. 272–3, note.) The statement given in the text is that held by Lockyer, Hind and others. The terms, it

year 1250 A. D., the earth was in perihelion when it was at the winter solstice, December 21, instead of January 1st, as now; the spring quarter was therefore equal to the summer one, and the autumn quarter to the winter one, the former being the longer. In the year 6589 A. D., the earth will be in perihelion when it is at the vernal equinox; summer will then be equal to autumn and winter to spring, the former seasons being the longer. In the year 11928 A. D., the earth will be in perihelion when it is at the summer solstice: finally, in 17267 A. D., the cycle will be completed, and for the first time since the creation of man the autumnal equinox will coincide with the earth's perihelion.

PERMANENCE IN THE MIDST OF CHANGE.—We thus see that the ecliptic is constantly modifying its elliptical shape; that the orbit of the earth slowly oscillates upward and downward; that the north pole steadily turns its long index-finger over a dial that marks 26,000 years; that the earth, accurately poised in space, yet gently nods and bows to the attraction of sun and moon. Thus changes are continually taking place that would ultimately entirely reverse the order of nature. But each of these has its bounds, beyond which it cannot pass. The promise made to man after the Deluge, is that "while the earth remaineth, seed-time and harvest, and cold and heat, and summer and winter, and

---

should be noticed, are applied to the real position of the earth and not the apparent position of the sun. The dates are those given by Chambers in his Descriptive Astronomy.

day and night shall not cease." The modern discoveries of astronomy prove conclusively that the seasons are to be permanent; that the Creator, amid all these transitions, has ordained the means of carrying out His promise through all time.

REFRACTION.—The atmosphere extends above the earth about 500 miles. Near the surface it is dense, while in the upper regions it is exceedingly rare. The rays of light from the heavenly bodies

Fig. 36.

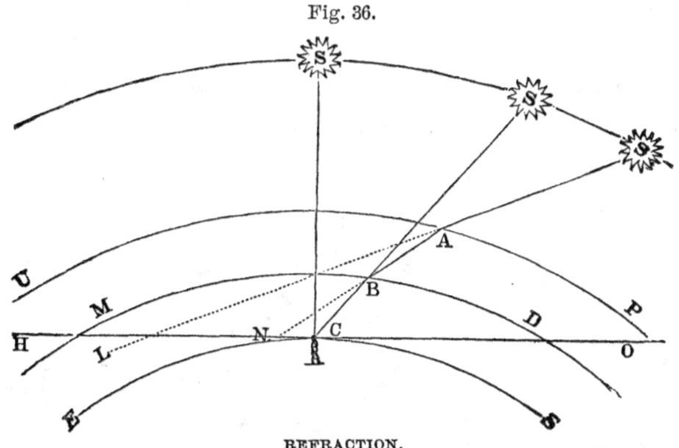

REFRACTION.

passing through these different layers are turned downward toward a perpendicular more and more as the density increases. According to a well-known law of optics, if the ray of light from a star were bent in fifty directions before entering the eye, the star would nevertheless appear to be in the line of the one nearest the eye. The effect of this is, that the apparent place of a heavenly body is higher than the true place. This is illustrated in Fig. 36. The sun at S, were it not for the atmosphere, would send a direct ray to L. Instead, the ray at A is refracted downward, and would then enter the eye at N; passing, however, through a layer of a different density, at B it is again bent, and meets the eye of the observer at C. He sees the sun, not in the direction of the curved line CBAS, but that of the straight line CBS.

The amount of refraction varies with the temperature, moisture, and other conditions of the atmosphere. It is zero for a body in the zenith, and increases gradually toward the horizon (as the thickness of the intervening atmosphere increases), where it is about 33′.

Fig. 37.

Change of place and appearance of the sun and moon.—The sun may be really below the horizon, and yet seem to be above it. For example, on April 20, 1837, the moon was eclipsed before the sun had set. The mean diameter of both the sun and moon being rather less than 33′, it follows that when we see the lower edge of either of these luminaries apparently just touching the horizon, in reality the whole disk is completely *below* it, and would be altogether hidden were it not for the effect of refraction. The day is consequently materially lengthened.

The sun and moon often appear *flattened* when near the horizon. This is easily accounted for on the principle just stated. The rays from the lower edge pass through a denser layer of the atmosphere, and are therefore refracted about 4′ more than those from the upper edge : the effect of this is to make the vertical diameter appear about 4′ less than the horizontal, and so distort the figure of the disk into an oval shape.

The sun and moon often appear larger when near the horizon than when high in the sky. This is not caused by refraction, but is a mere error of judgment. At the horizon we compare them with various terrestrial objects which lie between them and us, while aloft we have no association to guide us, and we are led to underrate their size. On looking at them through a tube, the illusion disappears. The moon should naturally appear largest when at a great altitude, as it is then at a less distance from us.

The dim and hazy appearance of the heavenly bodies when near the horizon is caused not only by the rays of light having to pass through a larger space in the atmosphere, but also by their traversing the lower and denser part. The intensity of the solar light is so greatly diminished by passing through the lower strata, that we are enabled to look upon the sun at that time without being dazzled by his brilliant beams.

Twilight.—The glow of light after sunset and before sunrise, which we term *twilight,* is caused by the refraction and reflection of the sun's rays by the atmosphere. For a time after the sun has truly set, the refracted rays continue to reach the earth; but when these have ceased, he still continues to illuminate the clouds and upper strata of the air, just as he may be seen shining on the summits of lofty mountains long after he has disappeared from the view of the inhabitants of the plains below. The air and clouds thus illuminated reflect back part of the light to the earth. As the sun sinks lower, less light reaches us until reflection ceases and night ensues. The same thing occurs before sunrise, only in reverse order. The duration of twilight is usually reckoned to last until the sun's depression below the horizon amounts to 18°; this, however, varies with the latitude, seasons, and condition of the atmosphere. Strictly speaking, in the latitude of Greenwich there is no true night for a month before and after the summer solstice, but constant twilight from sunset to sunrise. The sun is then near the Tropic of Cancer, and does not descend so much as 18° below the horizon during the entire night. The twilight is shortest at the equator and longest toward the poles, where the night of six months is shortened by an evening twilight of about fifty days and a morning one of equal length.

Diffused light.—The diffused light of day is produced in the same manner as that of twilight. The atmosphere reflects and scatters the sunlight in every direction. Were it not for this, no object would be visible to us out of direct sunshine; every shadow of a passing cloud would be pitchy darkness; the stars would be visible all day; no window would admit light except as the sun shone directly through it, and a man would require a lantern to go around his house at noon. This is illustrated very clearly in the rarified atmosphere of elevated regions, as on Mont Blanc, where it is said the glare of the direct sunlight is almost insupportable; the darkness of the shadows is deeper and denser; all nice shading and coloring disappear; the sky has a blackish hue, and the stars are seen at midday. The blue light reflected to our eyes from the atmosphere above us, or more probably from the vapor in the air, produces the optical delusion we call the sky. Were it not for this, every time we cast our eyes upward we should feel like one gazing over a dizzy precipice; while now the crystal dome of blue smiles down upon us so lovingly and beautifully that we call it heaven.

Aberration of Light.—We have seen that the places of the heavenly bodies are apparently changed by refraction. Besides this, there is another change due to the motion of light, combined with the motion of the earth in its orbit. For example: the mean distance of the earth from the sun is ninety-one and a half millions of miles, and since light travels 183,000 miles per second, it follows that the time occupied by a ray of light in reaching us from the sun is about 8½ min.; so that, in point of fact, when we look at the sun (1), we do not see it as it is, but as it was 8½ min. since. If our globe were at rest, this would be well enough, but since the earth is in motion, when the ray enters our eye we are at some distance in advance of the position we occupied when it started. During the 8½ min. the earth has moved in its orbit nearly 20½″, so that (2) we never see that luminary in the place it occupies at the time of observation.

Illustration.—Suppose a ball let fall from a point P, above the horizontal line A B, and a tube, of which A is the lower extremity, placed to receive it. If the tube were fixed, the ball would strike it on the lower side; but if the tube were carried forward in the direction A B, with a velocity properly adjusted at every instant to that of the ball, while *preserving its inclination* to the horizon, so that when the ball, in its natural descent, reached B, the tube would have been carried into the position BQ, it is evident that the ball throughout its whole descent would be found in the tube; and a spectator referring to the tube the motion of the ball, and carried

Fig. 38.

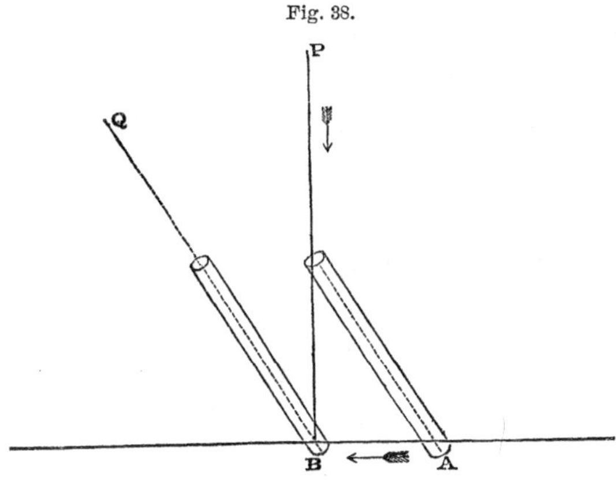

ABERRATION OF LIGHT.

along with the former, unconscious of its motion, would fancy that the ball had been moving in an inclined direction, and had come from Q. A very common illustration may be seen almost any rainy day. Choose a time when the air is still and the drops large. Then, if you stand still, you will see that the drops fall vertically; but if you walk forward, you will see the drops fall as if they were meeting you. If, however, you walk backward, you will observe that the drops fall as if they were coming from behind you. We thus see that the drops have an apparent as well as real motion.

The general effect of aberration of light is to cause each star to apparently describe a minute ellipse in the course of a year, the central point of which is the place the star would actually occupy were our globe at rest.

PARALLAX.—This is *the difference in the direction of an object as seen from two different places.* For a simple illustration of it, hold your finger before you

Fig. 39.

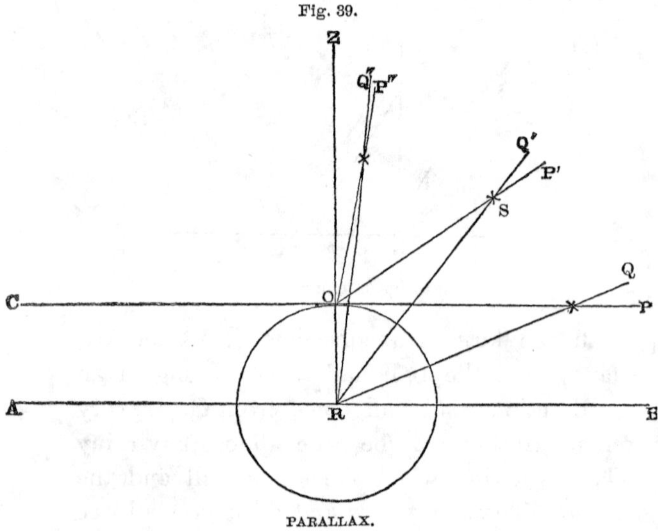

PARALLAX.

in front of the window. Upon looking at it with the left eye only, you will locate your finger at some point on the window; on looking with the right eye only, you will locate it at an entirely different point. Use your eyes alternately and quickly, and you will be astonished at the rate with which your finger will seem to change its place. Now, the difference in the direction of your finger as seen from the two eyes is *parallax.*

In astronomical calculations, the position of a body as seen from the earth's surface is called its *apparent* place, while that in which it would be seen from the centre of the earth is called its *true* place. Thus, in the cut, a star is seen by the observer at O in the direction OP; if it could be viewed from the centre R, its direction would be in the line RQ. It is therefore seen from O at a point in the heavens *below* its position in reference to R. From looking at the cut, we can see (1), that the parallax of a star near the horizon is greatest, while it decreases gradually until it disappears altogether at the zenith, since an observer at O, as well as one at R, would see the star Z directly overhead; and (2), that the nearer a body is to the earth the greater its parallax becomes. It has been agreed by astronomers, for the sake of uniformity in their calculations, to correct all observations so as to refer them to their true places as seen from the centre of the earth. Tables of parallax are constructed for this purpose. The question of parallax is also one of very great importance, because as soon as the parallax of a body is once accurately known, its distance, diameter, etc., can be readily determined. (See Celestial Measurements.)

Horizontal Parallax.—This is the parallax of

a body when at the horizon. It is, in fact, *the earth's semi-diameter as seen from the body.* In the figure, the parallax of the star S is the angle OSR, which is measured by the line OR—the semi-diameter of the earth. The *sun's horizontal parallax* (8.94″) is the angle subtended (measured) by the earth's semi-diameter as seen from that luminary. As the moon is nearest the earth, its horizontal parallax is the greatest of any of the heavenly bodies.

Annual Parallax.—The fixed stars are so distant from the earth that they exhibit no change of place when seen from different parts of the earth. The lines OS and RS are so long that they are apparently parallel, and it becomes impossible to discover the slightest inclination. Astronomers, therefore, instead of taking the earth's semi-diameter, or 4,000 miles, as the measuring tape, have adopted the plan of observing the position of the fixed stars at opposite points in the earth's orbit. This gives a change in place of 183,000,000 miles. The variation of position which the stars undergo at these remote points is called their *annual parallax.*

THE MOON.

New Moon, ●. First Quarter, ◑. Full Moon, ○. Last Quarter, ◐.

ITS MOTION IN SPACE.—The orbit of the moon, considering the earth as fixed, is an ellipse of which our planet occupies one of the foci. Its distance from the earth therefore, varies incessantly. At perigee it is 26,000 miles nearer than in apogee: the mean distance is about 238,000 miles. It would require a chain of thirty globes equal in size to the earth to reach the moon. An express-train would take about a year to accomplish the journey. The moon completes its revolution (*sidereal*) around the earth in about 27⅓ days; but, as the earth is constantly passing on in its own orbit around the sun, it requires over two days longer before it comes into the same position with respect to the sun and earth, thus completing its *synodic* revolution.

Fig. 40.

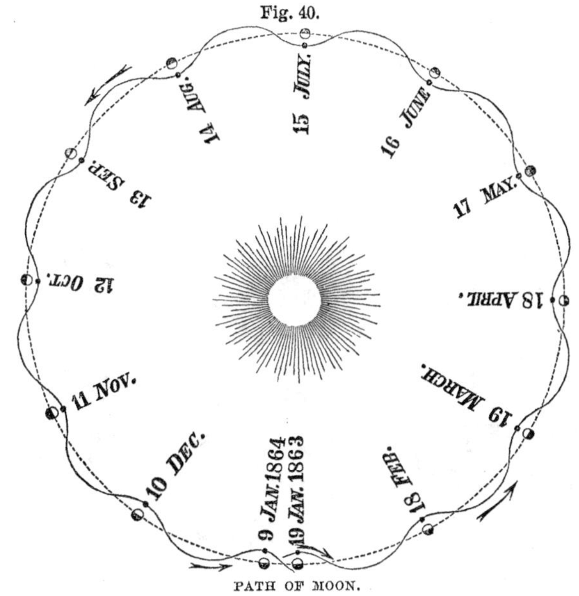

PATH OF MOON.

The real path of the moon is the result of its own proper motion and the onward movement of the earth. The two combined produce a wave-like curve that crosses the earth's path twice each month; this, owing to its small diameter compared with that of the ecliptic, is always concave toward the sun. As the moon constantly keeps the same side turned toward us, it follows that it must turn on its axis once each month.

DIMENSIONS.—Its diameter is about 2,160 miles. It would require fifty globes the size of the moon to equal the earth. Its apparent size varies with its distance; the mean is, however, about one half a

Fig. 41.

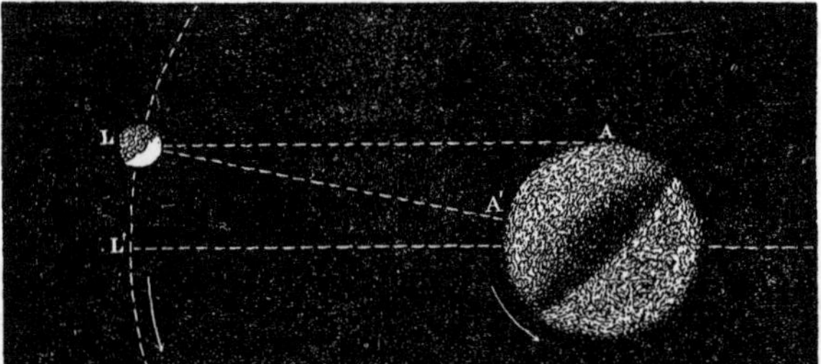

THE SIZE OF MOON AT HORIZON AND ZENITH.

degree, the same as that of the sun. It always appears larger than it really is, on account of its brightness. This is the effect of what is termed in optics *Irradiation.* To illustrate this principle, cut two circular pieces of the same size, one of black and the other of white paper. The white circle, when held in a bright light, will appear much larger than the black one. For the same reason it is often noticed that the crescent moon seems to be a part of a larger circle than the rest of the moon. As we have already said, the moon appears larger on the horizon than when high up in the sky. By an examination of the cut, it is easily seen that it is 4,000 miles nearer when on the zenith than when at the horizon. Besides these general variations in size, the moon varies in apparent size to different observers. Much amusement may be had in a large party or class by a comparison of its apparent magnitude. The estimates will differ from a small saucer to a wash-tub.

LIBRATIONS (*librans*, swinging).—While the moon presents the same hemisphere to us, there are three causes which enable us to see about 576 out of the 1,000 parts of its entire surface. (1.) The axis of the moon is inclined a little to its orbit, as also its orbit is inclined to the earth's orbit; so when its north pole leans alternately toward and from the earth, we see sometimes past its north, and sometimes past its south pole. This is called *libration in latitude.* (2.) The moon's rotation on its axis is always performed in the same time, while its movement along its orbit is variable; hence it happens that we occasionally see a little further around each *limb* (outer edge) than at other times. This is called *libration in longitude.* (3.) The size of the earth is so much greater than that of the moon, that an observer, by the rotation of the earth, or by going north or south, can see further around the limbs.

Light and Heat.—If the whole sky were covered with full moons, they would scarcely make daylight, since the brilliancy of the moon does not exceed $\frac{1}{300,000}$ that of the sun. That portion of the moon's surface which is exposed to the sun is supposed to be highly heated, possibly to the degree of boiling water, yet its rays impart no heat to us; indeed Prof. Tyndall considers them *rays of cold*. This is probably caused by the fact that our dense atmosphere absorbs all the heat, which in the higher regions produces the effect of scattering the clouds. It is a well-known fact that the nights are oftenest clear at full moon. (Herschel.)

Centre of Gravity.—It is thought that the centre of gravity of the moon is not exactly at its centre of magnitude, but nearly thirty-three miles beyond, and that the lighter half is toward us. If that be so, this side is equivalent to a mountain of that enormous height. We can easily see that if water and air exist upon the moon, they cannot remain on this hemisphere, but must be confined to the side which is forever hidden from our view.

Atmosphere of the Moon.—The existence of an atmosphere upon our satellite is at present an open question. If there be any, it must be extremely rarefied, perhaps as much so as that which is found in the vacuum obtained in the receiver of our best air-pumps.

Appearance of the Earth to Lunarians.—If there be any lunar inhabitants on the side toward us, the earth must present to them all the phases which their world exhibits to us, only in a reverse order. When we have a new moon, they have a *full earth*, a bright full-orbed moon fourteen times as large as ours. The lunar inhabitants upon the side opposite to us of course never see our earth, unless they take a journey to the regions from whence it is visible, to behold this wonderful spectacle. Those living near the limbs of the disk might, however, on account of the *librations*, get occasional glimpses of it near their horizon.

Fig. 42.

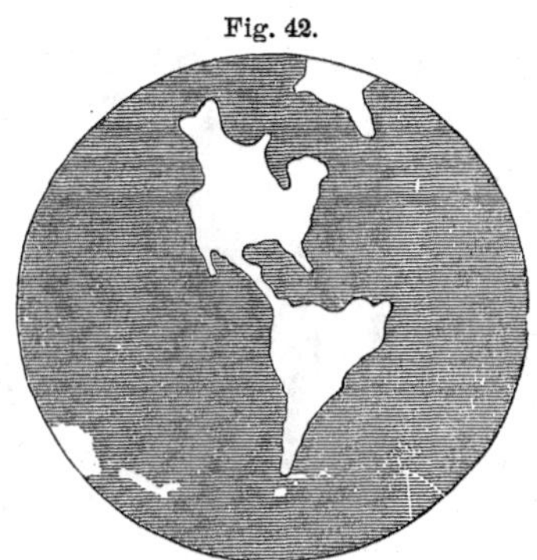

APPEARANCE OF EARTH AS SEEN FROM MOON.

The Earth-shine.—For a few days before and after new moon, we may distinguish the outline of the unillumined portion of the moon. In England, it is popularly known as "the old moon in the new moon's arms." This reflection of the earth's rays must serve to keep the lunar nights quite light, even in *new earth*.

Phases of the Moon.—The phases of the moon show conclusively that it is a dark body, which shines only by reflecting the light it receives from the sun. Let us compare its various appearances with the positions indicated in the figure.

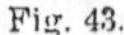
Fig. 43.

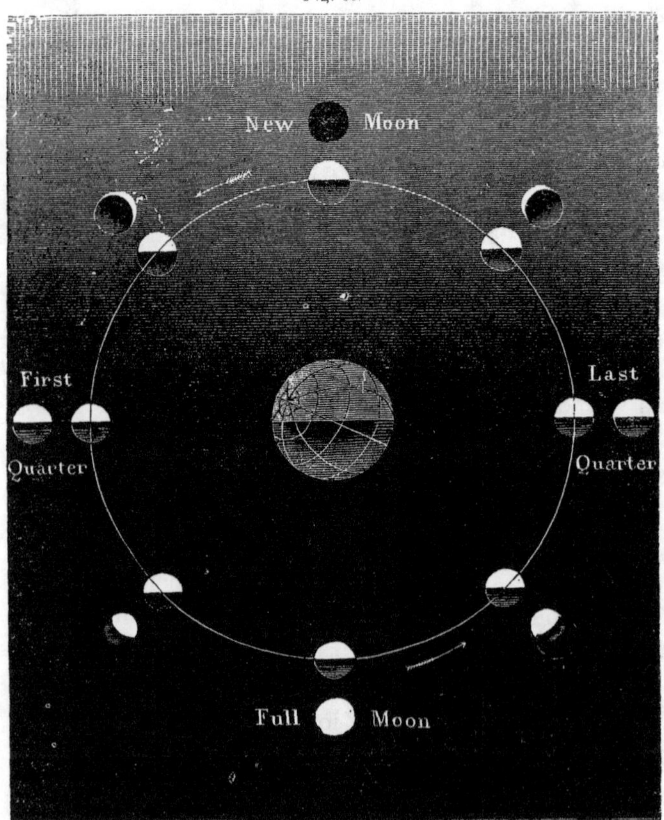

PHASES OF MOON.

We see it (1) as a delicate crescent in the west just after sunset, as it first emerges from the sun's rays at conjunction. It soon sets below the horizon. Half of its surface is illumined, but only a slender edge with its horns turned from the sun is visible to us. Each night the crescent broadens, the moon recedes about 13° further from the sun, and sets correspondingly later, until at quadrature half of the enlightened hemisphere is turned toward us, and the moon is said to be in her *first quarter*. Continuing her eastern progress round the earth, the moon (2) becomes *gibbous** in form, and, about the fifteenth day from new moon, reaches the point in the heavens directly opposite to that which the sun occupies. She is then in *opposition*, the whole of the illumined side is turned toward us, and we have a *full moon*. She is on the meridian at midnight, and so rises in the east as the sun sets in the west, and *vice versa*.

The moon (3) passing on in her orbit from opposition, presents phases reversed from those of the second quarter. The proportion of the illumined side visible to us gradually decreases; she becomes *gibbous* again; rises nearly an hour later each evening, and in the morning lingers high in the western sky after sunrise. She now comes into quadrature, and is in her *third quarter*.

From the third quarter the moon (4) turns her enlightened side from us and decreases to the crescent form again; as, however, the bright hemisphere

---

* *Gibbous* means less than a half and more than a quarter circle.

constantly faces the sun, the horns are pointed toward the west. She is now seen as a bright crescent in the eastern sky just before sunrise. At last the illumined side is completely turned from us, and the moon herself, coming into conjunction with the sun, is lost in his rays. To accomplish this journey through her orbit from new moon to new moon, has required 29½ days—a *lunar month.*

Moon runs high or low.—All have, doubtless, noticed that, in the long nights of winter, the full moon is high in the heavens, and continues a long time above the horizon; while in midsummer it is low, and remains a much shorter time above the horizon. This is a wise provision of Providence, which is seen yet more clearly in the arctic regions. There the moon, during the long summer day of six months, is above the horizon only for her first and fourth quarters, when her light is least; but during the tedious winter night of equal length, she is continually above the horizon for her second and third quarters. Thus in polar regions the moon is never full by day, but is always full every month in the night. We can easily understand these phenomena when we remember that the new moon is in the same quarter and the full moon in the opposite quarter of the heavens from the sun. Consequently, the moon always becomes full in the other solstice from that in which the sun is. When, therefore, the sun sinks very low in the southern sky the full moon rises high, and when the sun rises high the full moon sinks low.

Harvest Moon.—While the moon rises on the average 50 m. later each night, the exact time varies from less than half an hour to a full hour. Near the time of autumnal equinox the moon, at her full, rises about sunset a number of nights in succession. This gives rise to a series of brilliant moonlight evenings. It is the time of harvest in England, and hence has received the name of the Harvest Moon. Its return is celebrated as a festival among the peasantry. In the following month (October) the same occurrence takes place, and it is then termed the Hunter's Moon. The cause of this phenomenon lies in the fact that the moon's path is variously inclined to the horizon at different seasons of the year. When the equinoxes are in the horizon, it makes a very small angle with the horizon; whereas, when the solstitial points are in the horizon, the angle is far greater. In the former case, the moon moving eastward each day about 13°, will descend but little below the horizon, and so for several successive evenings will rise at about the same hour. In the latter, she will descend much further each day and thus will rise much later each night. The least possible variation in the hour of rising is 17 minutes—the greatest is 1 hour, 16 minutes.

In the figure, S represents the sun, E the earth, M the moon; C F the moon's path around the earth when the solstitial points are in the horizon—E D when the equinoxes are in the horizon; A M B S the horizon; M d = M b = 13°, the distance the moon moves each day. When passing along the path C F, the moon sinks below the horizon the distance $a b$, and when moving along the path E D, only the distance $c d$. It is obvious that before the moon can rise in the former case, the horizon must be depressed the distance a b, and in the latter only $c d$; and the moon will rise correspondingly later in the one and earlier in the other.

Fig. 44.

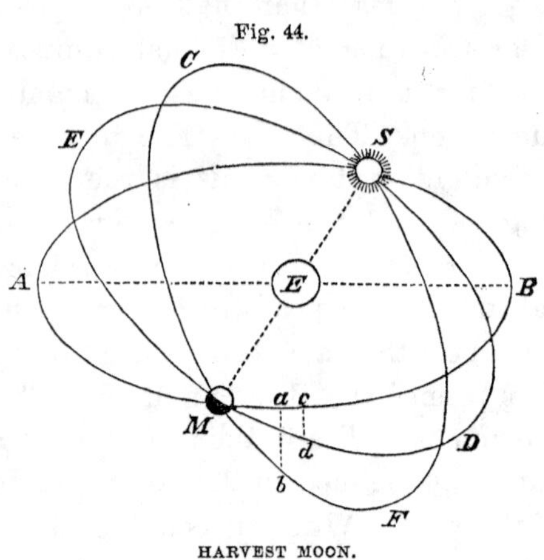

HARVEST MOON.

NODES.—The orbit of the moon is inclined to the ecliptic about 5°, the points where her path crosses it being termed *nodes*. The ascending node (☊) is the place where the moon crosses in coming above the ecliptic or toward the north star; the descending node (☋) is where it passes below the ecliptic. The imaginary line connecting these two points is called the "line of the nodes."

OCCULTATION.—The moon, in the course of her monthly journey round the earth, frequently passes in front of the stars or planets, which disappear on one side of her disk and reappear on the other. This is termed an occultation, and is of practical use in determining the difference of longitude between various places on the earth.

Lunar Seasons; Day and Night, etc.—As the moon's axis is so nearly perpendicular to her orbit, she cannot properly be said to have any change of seasons. During nearly fifteen of our days, the sun pours down its rays unmitigated by any atmosphere to temper them. To this long, torrid day succeeds a night of equal length and polar cold. How strange the lunar appearance would be to us! The disk of the sun seems sharp and distinct. The sky is black and overspread with stars even at midday. There is no twilight, for the sun bursts instantly into day, and after a fortnight's glare, as suddenly gives place to night; no air to conduct sound, no clouds, no winds, no rainbow, no blue sky, no gorgeous tinting of the heavens at sunrise and sunset, no delicate shading, no soft blending of colors, but only sharp outlines of sun and shade.

What a bleak waste! A barren, voiceless desert! The nights, however, of the visible hemisphere must be brilliantly illuminated by the earth, while its phases "serve well as a clock—a dial all but fixed in the same part of the heavens, like an immense lamp, behind which the stars slowly defile along the black sky."

Telescopic Features.—The lunar landscape is yet more wonderful than its other physical features.

Fig. 45.

IDEAL LANDSCAPE OF THE MOON.

Even with the naked eye we see on its surface bright spots—the summits of lofty mountains, gilded by the first rays of the sun—and darker portions, low plains yet lying in comparative shadow. The telescope reveals to us a region torn and shattered by fearful, though now extinct* volcanic action. Everywhere the crust is pierced by craters, whose irregular edges and rents testify to the convulsions our satellite has undergone at some past time.

Mountains.—The heights of more than 1,000 of these lunar mountains have been measured, some of which exceed 20,000 feet. The shadows of the mountains, as the sun's rays strike them obliquely, are as distinctly perceived as that of an upright staff when placed opposite the sun. Some of these are insulated peaks that shoot up solitary and alone from the centre of circular plains; others are mountain ranges extending hundreds of miles. Most of the lunar elevations have received names of men distinguished in science. Thus we find Plato, Aristarchus, Copernicus, Kepler, and Newton, associated however with the Apennines, Carpathians, etc.

Gray plains or seas.—These are analogous to our prairies. They were formerly supposed to be sheets of water, but have more recently been found to ex-

---

* Several distinguished astronomers assert, however, that the crater Linnæus has undergone of late certain marked changes. Its sides seem to have fallen in, and the interior to have become filled up, as if by a new eruption. It is said to present an appearance similar to that of the Sea of Serenity.

Fig. 46.

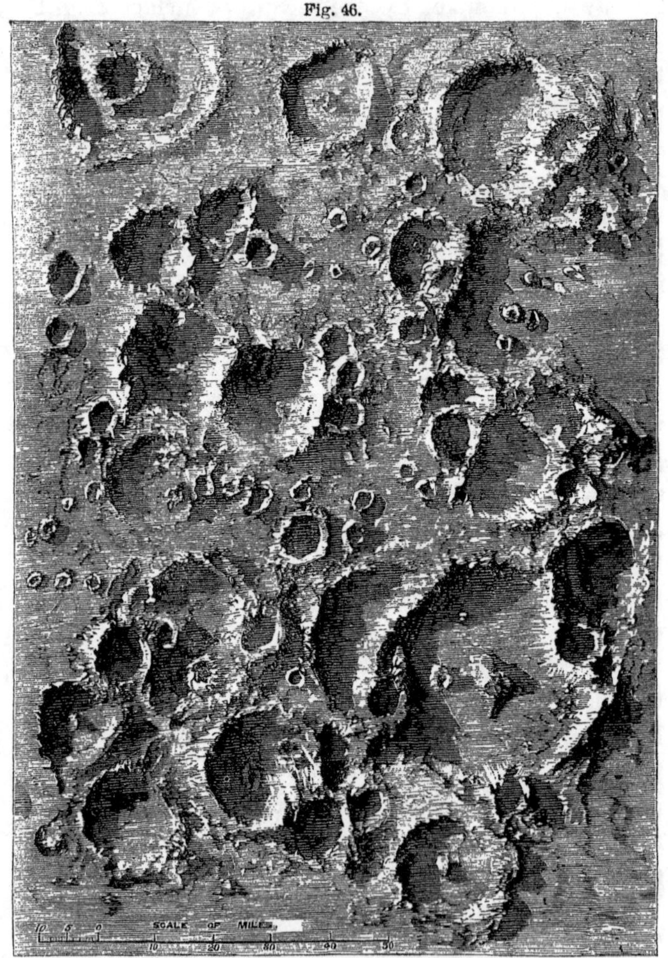

TELESCOPIC VIEW OF THE MOON

hibit the uneven appearances of a plain, instead of the regular curve of bodies of water. The former names have been retained, and we find on lunar maps the "Sea of Tranquillity," the "Sea of Nectar," "Sea of Serenity," etc.

Rills, luminous bands.—The latter are long bright streaks, irregular in outline and extent, which radiate in every direction from Tycho, Kepler, and other mountains; the former are similar, but are sunken, and have sloping sides, and were at first thought to be ancient river-beds. Their exact nature is yet a mystery.

Craters.—These constitute by far the most curious feature of the lunar landscape. They are of volcanic origin, and usually consist of a cup-like basin, with a conical elevation in the centre. Some of the craters have a diameter of over 100 miles. They are great *walled plains*, sunk so far behind huge volcanic ramparts, that the lofty wall which surrounds an observer at the centre would be beyond his horizon. Other craters are deep and narrow,—as Newton, which is said to be about four miles in depth,—so that neither earth nor sun is ever visible from a great part of the bottom. The appearance of these craters is strikingly shown in the accompanying view of the region to the southeast of Tycho. (Fig. 46.)

ECLIPSES.

Eclipse of the Sun.—If the moon should pass through either node at or near the time of conjunction or *new moon*, she would necessarily come between the earth and the sun, for the three bodies are then in the same straight line. This would cause

Fig. 47.

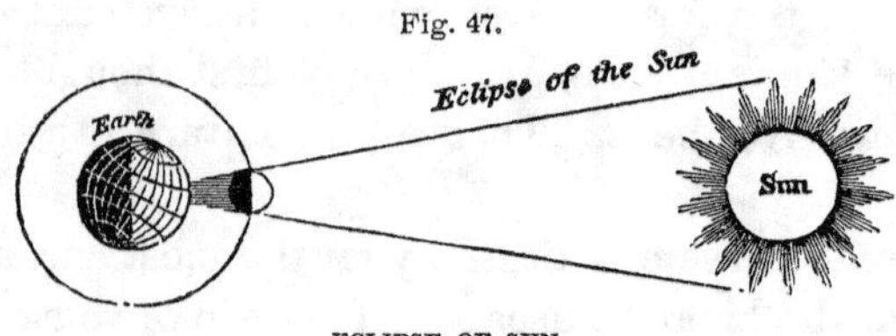

ECLIPSE OF SUN.

an eclipse of the sun. If the moon's orbit were in the same plane as the ecliptic, an eclipse of the sun would occur at every new moon; but as the orbit is inclined, it can occur only at or near a node.

The eclipse may be partial, total, or annular.—In Fig. 48, we see where the dark shadow (*umbra*) of

Fig. 48

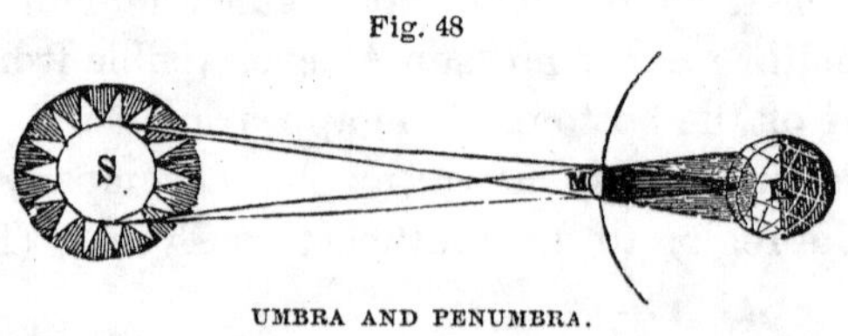

UMBRA AND PENUMBRA.

the moon falls on the earth and obscures the entire body of the sun. To the persons within that region there is a *total eclipse;* the breadth of this space is not large, averaging only 140 miles. Beyond this umbra there is a lighter shadow, *penumbra* (*pene,* almost--*umbra,* a shadow), where only a portion of the sun's disk is obscured. Within this region there is a *partial eclipse.* To those persons living north of the equator and of the umbra, the eclipse passes over the lower limb of the sun; to those south of the umbra, it passes over the upper limb.* When the eclipse occurs exactly at the node, it is said to be *central.* If the eclipse takes place when the moon is at apogee, or furthest from the earth, her apparent diameter is less than that of the sun; as a consequence, her disk does not cover the disk of the sun, and the visible portions of that luminary appear in the form of a ring (*annulus*); hence there is an *annular eclipse* in all those places comprised within the limits of the cone of shadow prolonged to the earth.

General facts concerning a solar eclipse.—The following data may perhaps guide in better understanding the phenomena of solar eclipses.

(1.) The moon must be new.

(2.) She must be at or near a node.

(3.) When her distance from the earth is less than the length of her shadow, the eclipse will be total or partial.

(4.) When her distance is greater than the length of her shadow, the eclipse will be annular or partial.

(5.) There can be no eclipse at those places where the sun himself is invisible during the time.

* South of the equator the reverse of these phenomena would happen.

(6.) An eclipse is not visible over the whole illumined side of the earth. As the moon's diameter is so much less than that of the earth, her cone of shadow is too small to enshroud the entire globe, so that the region in which it is total cannot exceed 180 miles in breadth. As, however, the earth is constantly revolving on its axis during the duration of the eclipse, the shadow may travel over a large surface of territory.

(7.) If the moon's shadow fall upon the earth when she is just nearing her ascending node, it will

Fig. 49

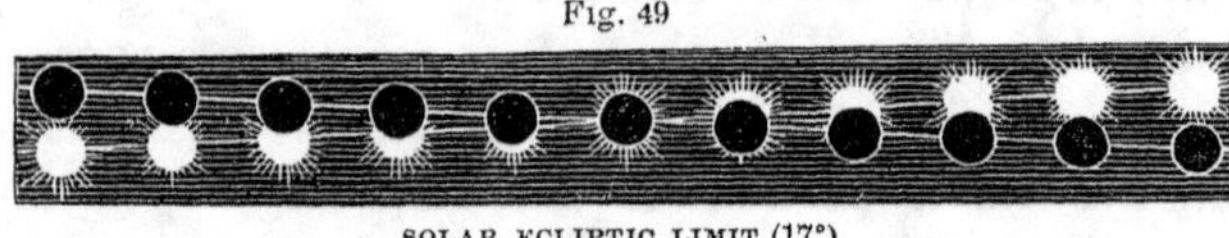

SOLAR ECLIPTIC LIMIT (17°).

only sweep across the south polar regions: if when nearing her descending node, it will graze the earth near the north pole. The nearer a node the conjunction occurs, the nearer the equatorial regions the shadow will strike.

(8.) At the equator, the longest possible duration of a total solar eclipse is only about eight minutes, and of an annular, twelve minutes. One reason of the greater length of the latter is, that then the moon is in apogee, when it always moves slower than when in perigee. The duration of total obscuration is greatest when the moon is in perigee and the sun in apogee; for then the apparent size of the moon is greatest and that of the sun least. We see from this that the relative situation of the moon and sun decides the length and kind of the eclipse.

(9.) There cannot be more than five nor less than two solar eclipses per year. A total or an annular eclipse is exceedingly rare. For instance, there has not been a total eclipse visible at London since 1715, and previous to that, there had been none visible for five and a half centuries.

(10.) A solar eclipse comes on the western limb or edge of the sun and passes off on the eastern.

(11.) The disk of the sun and moon is divided into twelve digits, and the amount of the eclipse is estimated by the number of digits which it covers. Thus an eclipse of six digits is one in which half the diameter of the disk is concealed.

Curious phenomena.—Various singular appearances always attend a total eclipse. Around the sun is seen a beautiful corona or halo of light, like that which painters give to the head of the Virgin Mary. Flames of a blood-red color play around the disk of the moon, and when only a mere crescent of the sun is

Fig. 50.

ECLIPSE OF 1858

visible, it seems to resolve itself into bright spots interspersed with dark spaces, having the appearance of a string of bright beads (Baily's Beads.)

Fig. 51.

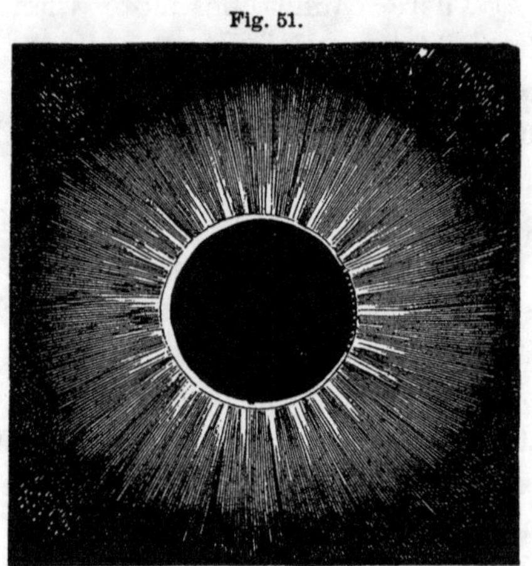

ANNULAR ECLIPSE OF 1836 SHOWING BAILY'S BEADS.

Attendant circumstances of a total eclipse.—These are of a peculiarly impressive character. The darkness is so intense that the brighter stars and planets are seen, birds cease their songs and fly to their nests, flowers close, and the face of nature assumes an unearthly cadaverous hue, while a sudden fall of the temperature causes the air to feel damp, and the grass wet as if from excessive dew. Orange, yellow, and copper tints give every object a strange appearance, and startle even the most indifferent. The ancients regarded a total eclipse with feelings of indescribable terror, as an indication of the anger of an offended Deity, or the presage of some impending calamity. Even now, when the causes are fully understood, and the time of the eclipse can be predicted within the fraction of a second, the change from broad daylight to instantaneous gloom is overwhelming, and inspires with awe even the most careless observer.

Curious custom among the Hindoos.—Among the Hindoos a singular custom is said to exist. When, during a solar eclipse, the black disk of our satellite begins slowly to advance over the sun, the natives believe that some terrific monster is gradually devouring it. Thereupon they beat gongs, and rend the air with most discordant screams of terror and shouts of vengeance. For a time their frantic efforts seem futile and the eclipse still progresses. At length, however, the increasing uproar reaches the voracious monster; he appears to pause, and then, like a fish rejecting a nearly swallowed bait, gradually disgorges the fiery mouthful. When the sun is quite clear of the great dragon's mouth, a shout of joy is raised, and the poor natives disperse, extremely self-satisfied on account of having so successfully relieved their deity from his late peril.

THE SAROS.—The nodes of the moon's orbit are constantly moving backward. They complete a revolution around the ecliptic in about eighteen and a half years. Now the moon makes 223 synodic revolutions in 18 yr. 10 da.; the sun makes 19 revolutions with regard to the lunar nodes in about the same time. Hence, in that period the sun and moon and the nodes will be in nearly the same relative position. If, then, we reckon 18 yr. 10 da. from any eclipse, we shall find the time of its repetition. This method was discovered, it is said, by the Chaldeans. The ancients were enabled, by means of it, to predict eclipses, but it is considered too rough by modern astronomers: eclipses are now foretold centuries in advance, true to a second. In this manner historical incidents are verified, and their exact dates determined.

METONIC CYCLE.—The Metonic Cycle (sometimes confounded with the Saros) was not used for foretelling eclipses, but for the purpose of ascertaining the *age of the moon* at any given period. It consists of nineteen tropical years,* during which time there are exactly 235 new moons; so that, at the end of this period, the new moons will recur at seasons of the year exactly corresponding to those of the preceding cycle. By registering, therefore, the exact days of any cycle at which the new or full moons occur, such a calendar shows on what days these events will occur in succeeding cycles. Since the regulation of games, feasts, and fasts has been made very extensively, both in ancient and modern times, according to new or full moons, such a calendar becomes very convenient for finding the day on which the new or full moon required takes place. Thus if a festival were decreed to be held in any given year on the day of the first full moon after the vernal equinox: find what year it is of the lunar cycle, then refer to the corresponding year of

---

* A tropical year is the interval between two successive returns of the sun to the vernal equinox.

the preceding cycle, and the day will be the same as it was then. The *Golden Number*, a term still used in our almanacs, denotes the year of the lunar cycle. Seven is the golden number for 1868.

ECLIPSE OF THE MOON.—This is caused by the passing of the moon into the shadow of the earth,

Fig. 52.

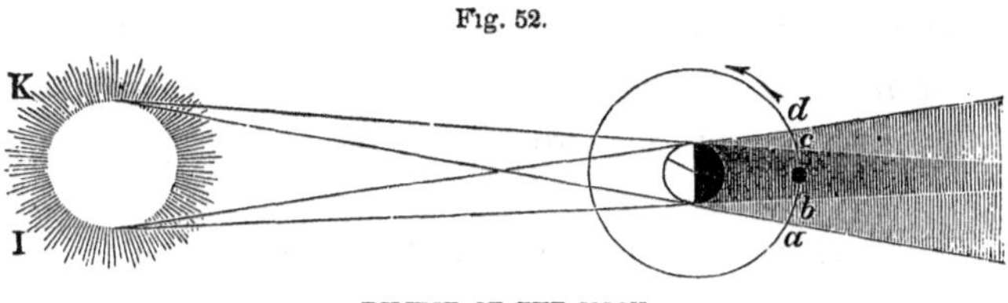

ECLIPSE OF THE MOON.

and hence can take place only at full moon—*opposition.* As the moon's orbit is inclined to the ecliptic, her path is partly above and partly below the earth's shadow; thus an eclipse of the moon can take place only at or near one of the nodes. In the figure, the *umbra* is represented by the space between the lines K*c* and I*b*; outside of this is the *penumbra,* where the earth cuts off the light of only a portion of the sun. The moon enters the *penumbra* of the earth at *a*,—this is termed her *first contact with the penumbra;* next she encounters the dark shadow of the earth at *b*,—this is called the *first contact with the umbra;* she then emerges from the umbra at *c*,—which is called the *second contact with the umbra;* finally, she touches the outer edge of the penumbra at *d*,—*the second contact with the penumbra.* Since the earth is so much larger than the moon, the eclipse can never be *annular*, as, however, the eclipse may occur a little above or below the node, the moon may only partly enter the earth's shadow, either on its upper or lower limb. From the first to last contact with the penumbra, five hours and a half may elapse. Total eclipses of the moon are rarer events than those of the sun, since the lunar ecliptic limit is only about 12° ; yet they are more frequently seen by us, (1) because each one is visible over the entire unillumined hemisphere of the earth, and also (2) because by the diurnal rotation during the long duration of the eclipse, large areas may be brought within its limits. So it will happen that while the inhabitants of one district witness the eclipse throughout its continuance, those of other regions merely see its beginning, and others only its termination. The moon does not completely disappear even in total eclipses. The cause of this fact lies in the refraction of the solar rays in traversing the lower strata of the earth's atmosphere ; they are analyzed, and purple our moon with the tints of sunset. The amount of refraction and the color depend upon the state of the air at the time.

Historical Accounts of Eclipses.—The earliest account of an eclipse on record is in the Chinese annals ; it is thought to be the solar eclipse of October 13, 2127 B. C. On May 28, 584 B. C., one occurred which was predicted by Thales, as we have before mentioned. In the writings of the early English chroniclers are numerous passages relating to eclipses. William of Malmesbury thus refers to that of August 2, 1133, which was considered a presage of calamity to Henry I.: "The elements manifested their sorrows at this great man's last departure. For the sun on that day, at the 6th hour, shrouded his glorious face, as the poets say, in hideous darkness, agitating the hearts of men by an eclipse; and on the 6th day of the week, early in the morning, there was so great an earthquake, that the ground appeared absolutely to sink down; an horrid noise being first heard beneath the surface." The same writer, speaking of the total eclipse of March 20, 1140, says: "During this year, in Lent, on the 13th of the kalends of April, at the 9th hour of the 4th day of the week, there was an eclipse, throughout England, as I have heard. With us, indeed, and with all our neighbours, the obscuration of the Sun also was so remarkable, that persons sitting at table, as it then happened almost every where, for it was Lent, at first feared that Chaos was come again: afterwards learning the cause, they went out and beheld the stars around the Sun. It was thought and said by many, not untruly, that the king [Stephen] would not continue a year in the government." Columbus made use of an eclipse of the moon, which took place March 1, 1504, to relieve his fleet, which was in great distress from want of supplies. As a punishment to the islanders of Jamaica, who refused to assist him, he threatened to deprive them of the light of the moon. At first they were indifferent to his threats, but "when the eclipse actually commenced, the barbarians vied with each other in the production of the necessary supplies for the Spanish fleet."

THE TIDES.

DESCRIPTION.—Twice a day, at intervals of about twelve hours and twenty-five minutes, the water begins to set in from the ocean, beating the pebbles and the foot of the rocky shore, and dashing its spray high in air. For about six hours it climbs far up on the beach, flooding the low lands and transforming simple creeks into respectable rivers. The instant of *high-water* or *flood-tide* being reached, it begins to descend, and the *ebb* succeeds the *flow*. The water, however, falls somewhat slower than it rose.

CAUSE OF THE TIDES.—The tides are caused by a great wave, which, raised by the moon's attraction,

Fig. 53.

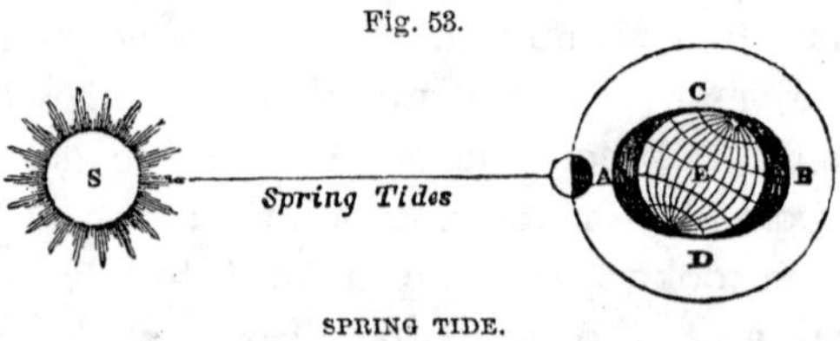

SPRING TIDE.

follows her in her course around the earth. The sun, also, aids somewhat in producing this effect; but as the moon is 400 times nearer the earth, her influence is far greater. As the waters are free to yield to the attraction of the moon, she draws them away from C and D and they become heaped up at A. The earth, being nearer the moon than the waters on the opposite side, is more strongly attracted, and so, being drawn away from them, they are left heaped up at B. As the result, high-water is produced at A by the water being pulled from the earth, and at B by the earth being pulled from the water. The influence of the moon is not instantaneous, but requires a little time to produce its full effect; hence high-water does not occur at any place when the moon is on the meridian, but a few hours after. As the moon rises about fifty minutes later each day, there is a corresponding difference in the time of high-water. While, however, the lunar tide-wave thus lags about fifty minutes every day, the solar tide occurs uniformly at the same time. They therefore steadily separate from each other. At one time they coincide, and high-water is the sum of lunar and solar tides; at other times, high-water of the solar tide and low-water of the lunar tide occur simultaneously, and high-water is the difference between the lunar and solar tides.

We should bear in mind the philosophical truth, that the tide in the open sea does not consist of a progressive movement of the water itself, but only of the form of the wave.

Causes that modify the tides.—At new and full moon (the *syzygies*) the sun acts with the moon (as in Fig.

53) in elevating the waters; this produces the highest or *Spring tide.* In quadrature (as in Fig. 54), the sun tends to diminish the height of the water: this is called *Neap-tide.* When the moon is in perigee her attraction is stronger; hence the flood-tide is higher and the ebb-tide lower than at other times. This re-

Fig. 54.

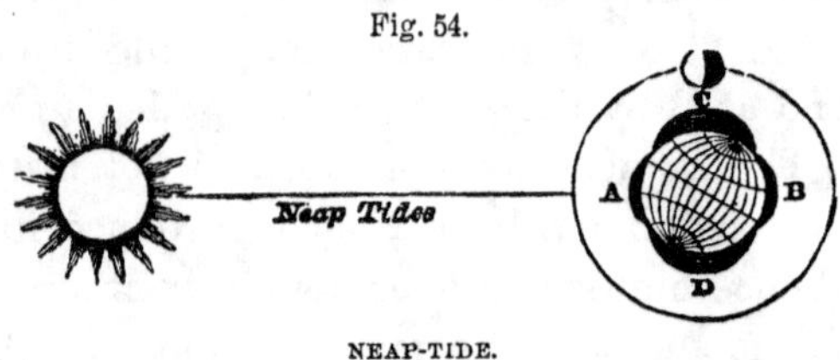

NEAP-TIDE.

mark applies also to the sun. The height of the tide also varies with the declination of the sun and moon, —the highest or equinoctial tides taking place at the equinoxes, if, when the sun is over the equator, the moon also happens to be very near it: the lowest occur at the solstices. The force and direction of the winds, the shape of the coast, and the depth of the sea wonderfully complicate the explanation of local tides.

Height of the tide at different places.—In the open sea the tide is hardly noticeable, the water sometimes rising not higher than a foot; but where the wave breaks on the shore, or is forced up into bays or narrow channels, it is very conspicuous. The difference between ebb and flood neap-tide at New York is over three feet, and that of spring tide over five feet; while at Boston it is nearly double this amount. A headland jutting out into the ocean will diminish the tide; as, for instance, off Cape Florida, where the average height is only one and a half feet. A deep bay opening up into the land like a funnel, will converge the wave, as at the Bay of Fundy, where it rolls in, a great roaring wall of water sixty feet high, frequently overtaking and sweeping off men and animals. The tide sets up against the current of rivers, and often entirely changes their character; for example, the Avon at Bristol is a mere shallow ditch, but at flood-tide it becomes a deep channel navigable by the largest Indiamen.

Differential effect.—The whole attraction of the moon is only $\frac{1}{120}$ that of the sun: yet her influence in producing the tides and precession is greater, because that depends not upon the *entire* attraction either exerts, but upon the *difference* between their attraction upon the earth's centre and upon the earth's nearest surface. For the moon, on account of her nearness, the proportion of the distance of these parts is treble that of the sun, and hence her greater effect.

MARS.

The god of war. Sign, ♂, shield and spear.

DESCRIPTION.—Passing outward in our survey of the solar system, we next meet with Mars. This is the first of the *superior* planets, and the one most like the earth. It appears to the naked eye as a bright red star, rarely scintillating, and shining with a steady light, which distinguishes it from the fixed stars. Its ruddy appearance has led to its being celebrated among all nations. The Jews gave it the appellation of "blazing," and it bore in other languages a similar name. At conjunction its apparent

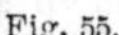

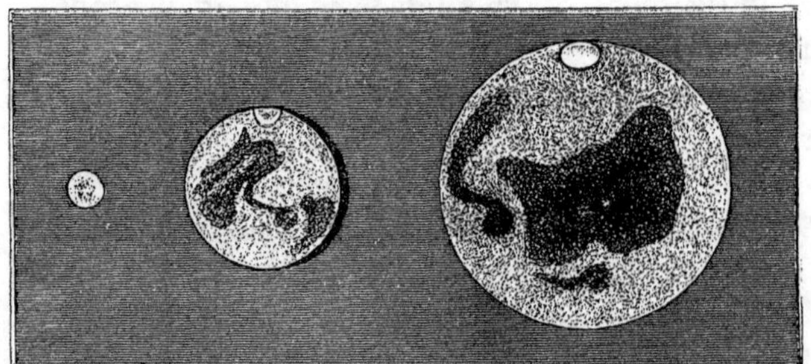

DIAMETER OF MARS AT EXTREME, LEAST, AND MEAN DISTANCES.

diameter is only about 4″; but once in two years it comes into opposition with the sun, when its diameter increases to 30″. At intervals of 8 yr. 7 mo. this occurs when the planet is also in perihelion and perigee. Mars then shines with a brilliancy rivalling that of Jupiter himself.

Motion in Space.—Mars revolves about the Sun at a mean distance of about 140,000,000 miles. Its orbit is sufficiently flattened to bring it at perihelion 26,000,000 miles nearer that luminary than when in aphelion. Its motion varies in different portions of its orbit, but the average velocity is about fifteen miles per second. The Martial day is about 40 min. longer than ours, and the year contains about 668 Martial days, equal to 687 terrestrial days (nearly two years).

DISTANCE FROM EARTH.—When in opposition, the distance of Mars is (like that of all the superior planets) the difference between the distance of the planet and that of the earth from the Sun: at conjunction it is the sum of these distances. If the orbits were circular, these distances would be the same at every revolution. The elliptical figure, however, occasions much variation. Thus, if it is in perihelion while the earth is in aphelion, the distance is 126,000,000 – 93,000,000 = 33,000,000 miles.

DIMENSIONS.—Its diameter is a little less than 5,000 miles. Its volume is about ¼ that of the earth, but as its density is only ½, it follows that its mass is only ⅛ of the terrestrial mass. A stone let fall on its surface would fall not quite five feet the first second. It is somewhat flattened at the poles, and bulges at the equator like our globe.

SEASONS.—The light and heat of the sun at Mars are less than one half that which we enjoy. Its axis is inclined about 28.7°, therefore its zones and seasons do not differ materially from our own: its days, also, as we have seen, are of nearly the same length Since, however, its year is equal to nearly two of our years, the seasons are lengthened in proportion. There must be a considerable difference between the temperature of its northern and southern hemispheres, as the former has its summer when 26,000,000 miles further from the sun than the latter: an increased length of 76 days may, however, be sufficient compensation. It has an atmosphere like our own, loaded with clouds. Mars has no moon. Its nights, therefore, are dark. Our own earth and moon must present in its evening sky a very beautiful pair of planets, showing all the phases which Mercury and Venus present to us, the two always remaining within one half the moon's apparent diameter of each other.

TELESCOPIC FEATURES.—Under the telescope, Mars exhibits slight phases, but by no means to the same

Fig. 56.

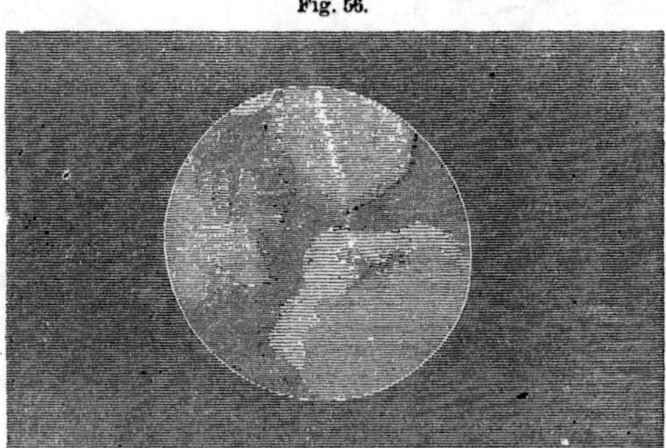

VIEW OF MARS.

extent as the inferior planets. Its surface appears covered with dusky patches, which are believed to be continents: these are of a dull red hue. Other portions, of a greenish tint, are considered to be bodies of water. The proportion of land to water on the earth is reversed in Mars. "*Here* every continent is an island; *there* every sea is a lake: but these, like our own continents, are chiefly confined to one hemisphere, so that the habitable area of the two globes may not differ so much as the size of the planets." The ruddy color of the planet is thought by Herschel to be due to an ochrey tinge in the soil; by others it is attributed to peculiarities of the atmosphere and clouds. Lambert suggests that it is the color of the vegetation, which, on Mars, may be red instead of green. There are constant changes going on in the brightness of the disk, owing, it is supposed, to the variation of the clouds of vapor in its atmosphere. No mountains have yet been discovered. In the vicinity of the poles are brilliant white spots, which are considered to be masses of snow. The "*snow zones*" apparently melt and recede with the return of summer in each hemisphere, and increase on the approach of winter. We can thus from the earth watch the formation of polar ice and the fall of snow—in fact, all the vicissitudes of the seasons on the surface of a neighboring planet.

THE MINOR PLANETS.

DISCOVERY.—Beyond Mars there is a wide interval which until the present century was not filled. The bold, imaginative Kepler conjectured that there was a planet in this space. This supposition was corroborated by Titius's discovery of what has since been known as Bode's law.

Take the numbers 0, 3, 6, 12, 24, 48, 96, 192, 384, each of which, after the second, is double the preceding one. If we add 4 to each of these numbers, we form a new series:

4, 7, 10, 16, 28, 52, 100, 196, 388.

At the time this law was discovered, these numbers represented very nearly the proportionate distance from the sun of the planets then known, taking the earth's distance as ten, except that there was a blank opposite 28.* This naturally led to inquiry, and a systematic effort to solve the mystery. On the 1st day of January, 1801, the nineteenth century was inaugurated by Piazzi's discovery of the small planet Ceres, at almost the exact distance necessary to fill the gap in Bode's series. This was soon followed by the announcement of other new planets, until (1870) there are one hundred and twelve, and a probability of many more. Indeed, Leverrier has calculated that there may be perhaps 150,000 in all.

* PLANETS.	True distance from ☉.	Distance by Bode's law.	PLANETS.	True distance from ☉.	Distance by Bode's law.
Vulcan........			Ceres........	27.66	28.00
Mercury	3.87	4.00	Jupiter......	52.03	52.00
Venus	7.23	7.00	Saturn.......	95.39	100.00
Earth	10.00	10.00	Uranus	191.82	196.00
Mars..........	15.23	16.00	Neptune.....	300.37	388.00

DESCRIPTION.—These minor worlds, or "pocket planets," as Herschel styled them, are extremely diminutive. The largest of them is Pallas, whose diameter is perhaps 600 miles. Those recently discovered are so small that it is difficult to decide which is the smallest. A French astronomer recently remarked concerning them, that a "good walker could easily make the tour of one in a day;" a prairie farmer would need to pre-empt a whole one for a flourishing cornfield. They all revolve about the sun in regular orbits, comprising a zone about 100,000,000 miles in width. Their paths are variously inclined to the ecliptic; Massilia's 41′, while that of Pallas rises 34°.

ORIGIN.—One theory concerning the origin of these small planets is, that they are the fragments of a large planet which, in a remote antiquity, has been shivered to pieces by some terrible catastrophe. "One fact seems above all others to confirm the idea of an intimate relation between these planets. It is this: if their orbits consisted of solid rings, they would be found so entangled that it would be possible, by taking up any one at random, to lift all the rest." Another theory is given under the "Nebular Hypothesis."

Names and signs.—Ceres, the first discovered, received the symbol ⚳, a sickle. This was appropriate, since that goddess was supposed to preside over harvests. Pallas, the second, named from the goddess of wisdom and scientific warfare, obtained the sign ⚴, the head of a spear. To Juno, the third planet, was assigned ⚵, a sceptre surmounted with a star, the emblem of the queen of heaven. An altar with fire upon it, ⚶, appropriately represented Vesta, the household goddess, whose sacred fire was kept burning continually. In this way names of goddesses and appropriate symbols were used to designate the minor planets which were earliest discovered. Since then a simple circle with the number inclosed has been adopted; thus ① represents Ceres—② is the sign of Pallas.

JUPITER.

The king of the gods. Sign ♃, a hieroglyphic representation of an eagle "the bird of Jove."

Description.—From the smallest members of the solar system we now pass at once to the largest planet—the colossal Jupiter. Its peculiar splendor and brilliancy distinguish it from the fixed stars, and vie even with the lustre of Venus. It is one of the five planets discovered in primitive ages. In those early times, Jupiter was supposed to be the cause of storm and tempest. Pliny thought that lightning owed its origin to this planet. An old almanac of 1368, foretelling the harmless condition of Jupiter for a certain month, says, "Jubit es hote and moyste and does weel til al thynges and noyes nothing."

Motion in Space.—Jupiter revolves about the sun at a mean distance of 475,000,000 miles. His orbit has much less eccentricity than those of the smaller planets. Were his path very elliptical, the attraction of the sun would be insufficient to bring him back from its extreme limit, and the huge unwieldy planet would plunge headlong into space. This careful fitting, whereby the plan is always modified to accomplish an end, is everywhere characteristic of nature, and is a continued revelation of its common Author. The revolution of Jupiter among the fixed stars is slow and majestic, comporting well with his vast dimensions and the dignity conferred by four attendant worlds. He advances through the zodiac at the rate of one constellation yearly; so that if we locate the planet now, a year hence we can find it equally advanced in the next sign. Yet slowly as he seems to travel through the heavens, he is bowling along through space at the enormous speed of 500 miles per minute. The Jovian day is only equal to about ten of our hours, while his year is lengthened to about 12 of our years, comprising near 10,000 of his days.

Distance from Earth.—Once in thirteen months Jupiter is in *opposition*, and his distance from the earth is measured by the difference of the distances of the two bodies from the sun. At the expiration of half this time he is in *conjunction*, and his distance from us is measured by the sum of these distances.

Dimensions.—Its diameter is about 88,000 miles, or one-tenth of the sun. Its volume is 1,400 times that of the earth, and much exceeds that of all the other planets combined. Seen at the distance of the moon, this immense globe would embrace 1,200 times the space of the full moon. Jupiter's density is only one-fifth that of the earth; moreover, its rapid rotation upon its axis, whereby a particle on the equator revolves with a velocity of 467 miles per minute against the earth's 17 miles per minute, must produce a powerful centrifugal force which materially diminishes the weight of all objects near its equator. Consequently a stone let fall on Jupiter would pass through but about thirty-nine feet the first second. As a result of this rapid rotation, the planet is one of the most flattened of any in the solar system—the equatorial diameter exceeding the polar by about 5,000 miles.

Fig. 57.

JUPITER.

Seasons.—As the axis of Jupiter is but slightly inclined from a perpendicular to the plane of its orbit, there is but little difference in the length of its days and nights, which are each of about five hours' duration. At the poles the sun is visible for nearly six years, and then remains set for the same length of time. The seasons also are but slightly varied. Summer reigns near the equator, while the temperate regions enjoy perpetual spring. The light and heat of the sun are only $\frac{1}{27}$ of that which we receive; yet peculiarities of soil or atmosphere may compensate this difference. The evening sky on Jupiter must be inexpressibly magnificent; besides the glittering stars which adorn our heavens, four moons, waxing and waning, each with its diverse phase, illuminate its night. All the starry exhibition sweeps through the sky in five hours.

TELESCOPIC FEATURES.—*Jupiter's moons.*—Under the telescope Jupiter presents a beautiful Copernican system in miniature. Four small stars—moons—are seen to accompany it in its twelve-yearly revolutions. From hour to hour their positions vary, and they seem to oscillate from one side to the other of the planet. At one time there will be two on each side, and again, three on one side; while the remaining star is left alone. They are also frequently found to disappear, one, two, or even three at a time, and, more rarely, all four at once. There are well-authenticated instances on record of their having been seen by the naked eye. Among others, the following singular case is mentioned. Wrangle, the celebrated Russian traveller, states, that when in Siberia, he once met a hunter, who said, pointing to Jupiter, "I have just seen that star swallow a small one and then vomit it up again." These moons are called by the ordinal numbers, reckoning outward from the planet. With an ordinary glass, there is nothing to distinguish them from small stars. The IIId, however, being the largest and brightest, will generally be identified easiest. The Ist satellite appears to the inhabitants of the planet almost as large as our moon to us; the IId and IIId about half as large. Their real size and density are indicated in the following table. It will be seen that the IVth is about the weight of cork, and the Ist and IId are still lighter.

SATELLITES OF JUPITER.

	Mean distance from Jupiter.	Diameter.	Density. Water as 1.	Sidereal period.		
				D.	H.	M.
I. Io..............	267,380	2,352 m.	.114	1	18	28
II. Europa	425,156	2,099 "	.171	3	13	4
III. Ganymede	678,393	3,436 "	.396	7	3	43
IV. Callisto.........	1,192,823	2,929 "	.222	19	16	32

It is noticeable that here are four satellites revolving about Jupiter, one of them larger than the planet Mercury, and each far surpassing in size the minor planets between Mars and Jupiter. The moons are not only thus distinguished by their various dimensions, but also by the variety of their color. The Ist and IId have a bluish tint, the IIId a yellow, and the IVth a reddish shade. The total space occupied by this miniature system is about two and a half million miles in diameter.

Eclipse of the moons.—Jupiter, like all celestial bodies not self-luminous, casts into space a cone of shade.

The Ist, IId, and IIId satellites revolve in orbits but very little inclined to the plane of the planet's orbit. During each revolution they pass

Fig. 58.

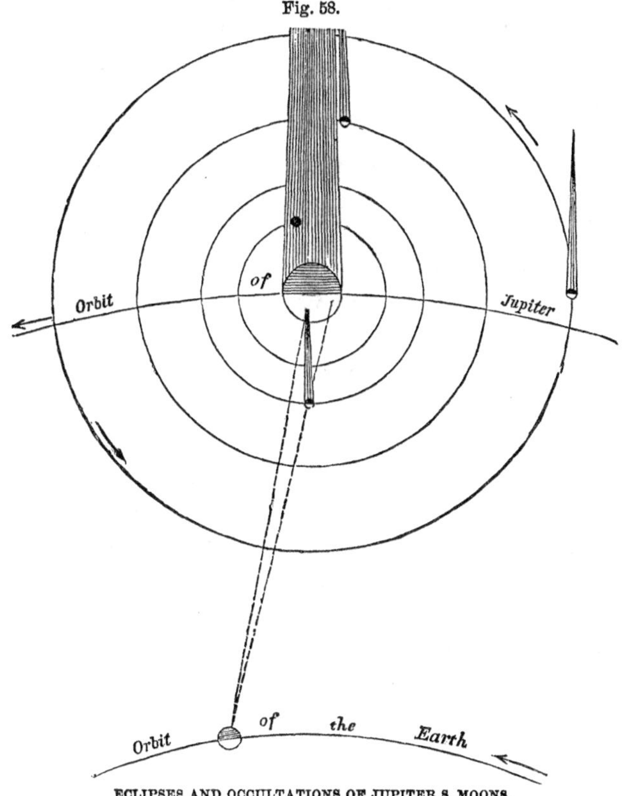

ECLIPSES AND OCCULTATIONS OF JUPITER S MOONS.

between the Sun and Jupiter, producing a solar eclipse; and also by passing through the shadow of the planet itself, cause to themselves an eclipse of the sun, and to Jupiter an eclipse of a moon. The IVth passes through a path more inclined, and therefore its eclipses are less frequent: instead of being fully eclipsed, it sometimes just grazes the shadow, as it were, and so its light is much diminished. Through a telescope we can distinctly watch the disappearance or *immersion* of the satellites in the planet's shadow, their reappearance or *emersion*, and also their *transits*, as a round black dot or shadow moving across the disk of Jupiter. In the cut, we see represented the various positions of the moons: the Ist is eclipsed; the IId is passing across the disk of the planet on which its shadow is also thrown; the IIId is just behind the planet, and so *occulted* or concealed, while it has not yet entered the shadow; the IVth is in view from the earth. These satellites revolve with great rapidity, as is necessary in order to overcome the superior attraction of the planet and prevent being drawn to its surface. The Ist goes through all its phases in $1\frac{3}{4}$ days, and the IVth in less than twenty days. A spectator on Jupiter might witness, during the Jovian year, 4,500 eclipses of the moon (moons), and about the same number of the sun.

Jupiter's belts.—These are dusky streaks of varying breadth and number, lying more or less parallel to the planet's equator, but terminating at a short distance from the edges of the disk. Between these a brighter, often rose-colored space, marks the equatorial regions. They are not permanent, but change sometimes very materially in the course of a few minutes. Occasionally only two or three broad belts are seen; at other times a dozen narrow ones appear. It is supposed that the planet is enveloped in dense masses of cloud, and that the belts are merely fissures, laying bare the solid body beneath. The parallel appearance is doubtless due to strong equatorial currents, analogous to our trade-winds.

Velocity of Light.—By an attentive examination of the eclipses of Jupiter's moons, Römer (a Danish astronomer, in 1617) was led to discover the progressive motion of light. Before him, it had been considered instantaneous. He noticed that the observed times of the eclipses were sometimes earlier and sometimes later than the calculated times, according as Jupiter was nearest or furthest from the earth. His investigations convinced him that it requires about 16½ min. for light to traverse the orbit of the earth. Römer's conclusion has since been verified by the phenomena of aberration of light. The velocity of light is about 183,000 miles per second. (See 14 Weeks in Philosophy, p. 189.)

SATURN.

The god of time. Sign ♄, an ancient scythe.

Description.—We now reach, in our outward journey from the sun, the most remote world known to the ancients. On account of its distance, it shines with a feeble but steady pale yellow light, which distinguishes it from the fixed stars. Its orbit is so vast that its movement among the constellations may be easily traced through one's lifetime. It requires two and a half years to pass through a single sign of the zodiac; hence, when once known, it may be easily found again. The earth leaves it at conjunction, makes a yearly revolution about the sun, comes to its starting point, and overtakes Saturn in about thirteen days thereafter.* On account of its slow, dreary pace, Saturn was chosen by the ancients as the symbol for lead. It is smaller than Jupiter, but much more gorgeously attended. Besides a retinue of eight satellites, it is surrounded by a system of rings, some shining with a golden light and others transparent—a spectacle which is as wonderful as it is unique.

Fig. 59.

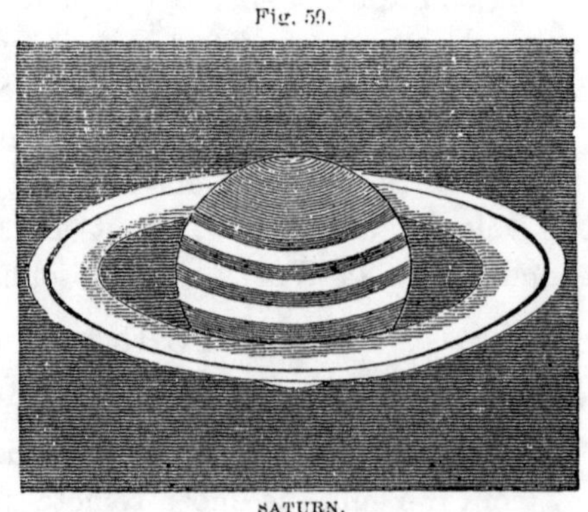

SATURN.

MOTION IN SPACE.— Saturn revolves about the sun at a mean distance of 872,000,000 miles. The eccentricity of its orbit is a trifle more than that of Jupiter,

* From this the year of Saturn may be determined. As 13 : 378 days :: Earth's year : Saturn's year = 30 yr. nearly

so that while it may at perihelion come fifty million miles nearer than its mean distance, at aphelion it swings off as much beyond. We can form some estimate of the size of its immense orbit, when we remember that it is moving along at the rate of 21,000 miles per hour, and yet as we look at it from night to night, we can scarcely detect any change of place. The Saturnian year is equal to about thirty of ours, and comprises nearly 25,000 Saturnian days, each of which is about ten and a half hours in length.

DISTANCE FROM EARTH.—This is found in the same manner as that of the other superior planets, being least in opposition and greatest at conjunction. As the earth and Saturn occupy different portions of their orbits, the distances between them at different times may vary 200,000,000 miles.

DIMENSIONS.—Its diameter is about 72,000 miles. Its volume is nearly 750 times that of the earth. Its density is very low indeed, being much less than that of water, and about the same as that of pine wood. The Saturnian force of gravity is therefore scarcely greater than the terrestrial, so that a stone falls toward the surface of that immense globe only about seventeen feet the first second.

SEASONS.—The light and heat of the sun at Saturn are only $\frac{1}{100}$ that which we receive. The axis of Saturn is inclined from a perpendicular to the plane of its orbit about 31°. The seasons therefore are similar to those on the earth, but on a larger scale. The sun climbs in summer about 8° higher above the horizon, and sinks correspondingly lower in winter. The tropics are 16° further apart, and the arctic and antarctic circles 8° further from the poles. Each of Saturn's seasons lasts more than seven of our years. There is about fifteen years interval between the autumn and spring equinoxes, and between the summer and winter solstices. For fifteen years the sun shines on the north pole, and a night of the same length envelops the south pole. The atmosphere is doubtless very dense, as the belts would seem to indicate.

TELESCOPIC FEATURES.—*Saturn's Rings.* Galileo first noticed something peculiar in the shape of Saturn. Through his imperfect telescope it seemed to have on each side a small planet like a supporter, to help old Saturn on his way. He therefore announced to his friend Kepler his curious discovery, that "Saturn is threefold." As the planet, however, approached its equinoxes, these attendants vanished altogether from his simple instrument. This was a great perplexity to Galileo, and he never solved the mystery. When the rings were afterward seen, their real form was not known. They were supposed to be a kind of *handle* attached to the planet, but for what purpose was not explained.

The series consists of three rings of unequal breadth, surrounding the planet at the equator. The exterior ring is separated from the middle one by a distinct break, while the interior one seems joined to the middle one. They differ in their brightness the exterior ring is of a grayish tint; the middle one is the most brilliant and is more luminous than Saturn itself; the interior is dusky and has a purple tinge. The exterior and middle rings are both opaque and cast on the planet a distinct shadow; while the interior one is so transparent that it appears upon the globe of Saturn as a dark band through which the surface of the planet is readily seen. The dimensions of the rings are given in the following table (Guillemin):

	Miles.
Diameter of exterior ring	173,500
Breadth of exterior ring	10,000
Diameter of middle ring	150,000
Breadth of middle ring	18,300
Distance between exterior and middle ring	1,750
Diameter of interior ring	113,400
Breadth of interior ring	9,000
Distance of interior ring from planet	10,150
Entire breadth of ring system	39,050
Thickness of rings not more than	100

The rings revolve around Saturn in about 10½ hours, in the same direction as the planet revolves on its axis. The globe of Saturn is not exactly at the centre of the rings. This fact, combined with the rotary motion, is essential to the stability of the rings, preventing them from being precipitated in an overwhelming ruin and devastation upon the body of the planet.

Phases of the rings.—The plane of the rings is inclined 28° to the ecliptic. In its revolution about the sun, the axis of Saturn remaining parallel to itself, the sun sometimes illumines the northern and sometimes the southern face of the rings. At Saturn's equinoxes the edge only receives the light, and the rings are invisible to us, except with the

Fig. 60.

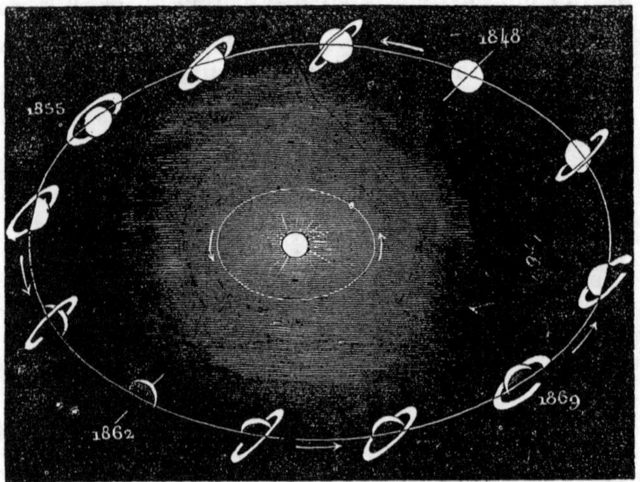

PHASES OF SATURN'S RINGS.

most powerful telescopes, and then only as a line of light. The body of the planet constantly cuts off the sun's rays from a portion of the rings, and also serves to conceal from our view some of the luminous part. By a careful study of the cut these various positions of the planet and rings, with the most favorable times for observation, may be understood.

Belts.—The surface of Saturn is traversed by dusky belts of a less distinct and definite appearance than those upon Jupiter. The equatorial regions are brighter than the other parts of the disk; the poles especially are less luminous.

SATELLITES.—Saturn has eight satellites, named—

1. Mimas.	3. Tethys.	5. Rhea.	7. Hyperion.
2. Enceladus.	4. Dione.	6. Titan.	8. Iapetus.

Iapetus is the largest of these, and in size exceeds Mars. Enceladus and Mimas are the faintest of twinklers, and can only be seen with a powerful telescope, and under most favorable circumstances. They were first detected by Herschel, "threading like pearls the silver line of light," to which the ring, then seen edgewise, was reduced,—advancing off it at either end, returning, and then hiding themselves behind the planet. The first four of these moons are nearer to Saturn than our moon to the earth, but Iapetus is nearly ten times as distant: so that the diameter of the Saturnian system is nearly four and a half million miles. The movements are extremely rapid. Mimas traverses a space equal to the diameter of our moon in two minutes, passing from new to full in twelve hours,—a little more than a Saturnian day.

SATURNIAN SCENERY.—The grandeur and magnificence of the scenery upon Saturn undoubtedly far surpass anything with which we are familiar. In the cut is given an ideal view of a landscape located upon the planet at a latitude of about 28°, taken about midnight. The rings form an immense arch, which spans the sky and sheds a soft radiance around; while to add to the strange beauty of the

Fig. 61.

IDEAL LANDSCAPE ON SATURN.

Saturnian night, eight moons in all their different phases, full, new, crescent, or gibbous, light up the starry vault.

URANUS.

"Heaven," the most ancient of the gods. Sign ♅; H, the initial letter of Herschel, with a planet suspended from the cross-bar.

DESCRIPTION.—On the 13th of March, 1781, between 10 and 11 P. M., Sir William Herschel was engaged in examining with his great telescope some stars in the constellation Gemini. One small star attracted his attention, which he accordingly observed with a higher magnifying power, when, unlike the effect produced on the fixed stars, its disk widened. Watching it for several nights, he detected its motion in space, and, mistaking its true character, announced the discovery of a new *comet.* A few months' examination revealed the error, and the new body was universally admitted to be a member of the solar system—new to us, but older perhaps than our own world. It is now known that Uranus had been previously observed by other astronomers. Indeed, Le Monier at Paris had watched it for twelve successive nights, but pronounced it a fixed star. Since he had also seen it on previous occasions, had he been an orderly observer, he would doubtless have detected its planetary character; but he was extremely careless, as may be inferred from the fact related by Arago, that he had been shown one of Le Monier's observations of this planet written on a paper bag which originally contained hair-powder purchased at a perfumer's. Uranus may be seen by a person of strong eyesight in a perfectly dark sky, if he previously knows its exact position among the stars. Its faintness is due to its great distance from the earth. Were it as near as the sun, it would appear twice as large as Jupiter.

MOTION IN SPACE.—Uranus revolves about the sun at a mean distance of 1,754,000,000 miles. Its year exceeds eighty-four of ours.

DIMENSIONS.—Its diameter is about 33,000 miles. It is lighter than water, having a density about equal to that of ice.

SEASONS.—We know little of the seasons of Uranus. Since its axis lies in the plane of its orbit, the sun winds in a spiral form around the whole planet. The light and heat are only $\frac{3}{1000}$ of that which we receive; the light is about the quantity which would be afforded by three hundred full moons. The inhabitants of Uranus can see Saturn, and perhaps Jupiter, but none of the planets within the orbit of the latter.

TELESCOPIC FEATURES.—No spots or belts have been discovered with any telescope yet made. The time of rotation and other features so familiar to us in the nearer planets, are unknown with regard to Uranus.

Satellites.—Uranus has four moons, of which little is known except the curious fact that their orbits are nearly perpendicular to the plane of the planet's orbit, and that their movements are retrograde—*i. e.*, in the same direction as the hands of a watch.

NEPTUNE.

The god of the sea. Sign ♆, his trident.

DESCRIPTION.—Neptune is the far-off sentinel at the very outposts of the solar system, being the most distant planet of which we have any knowledge. It is invisible to the naked eye, and appears in the telescope as a star of the eighth magnitude.

DISCOVERY.—For many years the motions of Uranus were such as to baffle the most perfect calculations. While far-distant Saturn came around to his place true to the minute and second, even after his journey of nearly thirty years, Uranus defied arithmetic, and refused to conform to the time set down for him on the heavenly dial.

At length it was suggested by several astronomers that there was another planet outside of its orbit, whose attraction produced these perturbations. So marked was this impression with Herschel, that he writes: "We see it as Columbus saw America from the shores of Spain. Its movements have been felt trembling along the far-reaching line of our analysis with a certainty not far inferior to ocular demonstration." Finally, two young mathematicians, Leverrier of Paris, and Adams of Cambridge, England, each unknown to the other, set themselves about the task of finding the place of this new planet. The problem was this: *Given the disturbances produced by the attraction of the unknown planet, to find its orbit and its place in the orbit.* Adams, after assiduous labor for nearly two years, completed his calculations and submitted them to Prof. Airy, the Astronomer Royal, in October, 1845. In the summer of 1846, Leverrier laid a paper before the Academy of Sciences in Paris, announcing the position of the unknown planet. Prof. Airy, hearing of this, was so impressed with the value of Adams's calculations, that he wrote to Prof. Challis, of Cambridge, to use his large telescope to search that quarter of the heavens. Prof. Challis did as requested, and saw a star which afterward proved to be the planet so anxiously sought for, although at that time he failed to ascertain its true character. On September 23d, of the same year, Leverrier wrote to Berlin, asking for assistance in searching for the planet. Dr. Galle, that same evening, turned the large telescope of the Observatory to the place indicated, and almost immediately detected a bright star not laid down in the maps. This proved to be the predicted planet, found within less than a degree of the spot described by Leverrier. Such is the history of one of the grandest achievements of the human mind. It stands as an ever fresh and assuring proof of the exactness of astronomical calculations, and the power of the intellect to understand the laws of the God of Nature.

Motion in Space.—Neptune revolves about the sun at a mean distance of about 2,750,000,000 of miles. The Neptunian year is equal to nearly 165 terrestrial ones. Its motion in its orbit is the slowest of any of the planets, since it is the most remote from the sun. The velocity decreases from Mercury, which moves at the rate of 105,000 miles per hour, to Neptune, whose rate is only 12,000 miles.

Dimensions.—Its diameter is about 37,000 miles. Its volume is nearly 100 times that of the earth. Its density is about that of Uranus, a little less than that of water.

Seasons.—As the inclination of its axis is unknown, nothing can be ascertained concerning its seasons. The sun gives to Neptune but $\frac{1}{1000}$ the light and heat which we receive.

Though at the extreme of the solar system, 2,650 millions of miles beyond us, the same heavens bend above, the same starry sky is seen by night—the Milky Way is no nearer to the eye, the fixed stars shine no more brightly. The planets, however, are all too near the sun to be seen, except Saturn and Uranus. The Neptunian astronomers, if there be any, are well situated for observing the orbits of comets, and for measuring the annual parallax of the stars, since they have an orbit of 5,500 million miles in diameter, and hence the angle must be 30 times as great as that which the terrestrial orbit affords.

TELESCOPIC FEATURES.—On account of the recentness of the discovery of this planet and its immense distance, nothing is known of its rotation or physical features.

Satellites.—Neptune has one moon, at nearly the same distance from it as our own moon is from the earth. The revolution of this about the planet, which is accomplished in about six days, has furnished the materials for calculating the mass of Neptune.

METEORS AND SHOOTING STARS.

DESCRIPTION.—All are familiar with those luminous bodies that flash through our atmosphere as if the stars were indeed falling from heaven. Different names have been applied to them, although the distinction is not very definite. (1) *Aërolites* are those

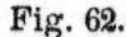

A METEOR WITH ITS TRAIN.

stony masses which fall to the earth. (2) *Shooting Stars* are those evanescent brilliant points that suddenly dart through the higher regions of the air, leaving a fiery train behind. (3) *Meteors* are luminous bodies which have a sensible diameter and a spherical form. They frequently pass over a great extent of country, and are seen for some seconds of time. Many leave behind a train of glowing sparks; others explode with reports like the discharge of artillery,—the pieces either continuing their course, or falling to the earth as aërolites. Some meteors, doubtless, after having favored us with a transient illumination, pass on into space; some are vaporized; while others are burned and the ashes and fragments fall to the ground.

Aërolites.—The fall of aërolites is frequently mentioned and well authenticated. Chinese records tell of one as long ago as in 616 B. C., which, in its fall, broke several chariots and killed ten men. A block of stone, equal to a full wagon-load, fell in the Hellespont, B. C. 465. By the ancients, these stones were held in great repute. The Emperor Jehangire, it is related, had a sword forged from a mass of meteoric iron which fell in the Punjab in 1620. In 1795, a mass was seen, by a ploughman, to descend toward the earth at a spot not far from where he was standing. It threw up the soil on every side, and penetrated some distance into the solid rock beneath. In 1807, a shower of stones, one weighing 200 lbs., fell at Weston, Connecticut. These aërolites are sometimes seen to plunge downward into the earth, and are found while yet glowing. A mass thus fell in

South America, which was estimated to weigh fifteen tons. When first discovered, it was so hot as to prevent all approach. Upon its cooling, many efforts were made, by some travellers who were present, to detach specimens, but its hardness was too great for any tools which they possessed. There is a mass of meteoric iron in Yale College cabinet, weighing 1,635 lbs.

Aërolites consist of elements which are familiar. The analysis of these stellar masses gives us names as commonplace as if they had known a far less romantic origin—oxygen, sulphur, phosphorus, iron, tin, copper: in all, nineteen elements have been found. This fact is interesting as revealing something of the chemistry of the region of space, concerning which we otherwise know nothing. The compounds, however, are very peculiar, so as to distinguish an aërolite from any terrestrial substance. For example, meteoric iron, a prominent constituent of aërolites, is an alloy that has never been found in terrestrial minerals.

METEORS.—The records of meteors are still more wonderful. It is related that at Crema, Italy, one day in the 15th century, the sky at noonday became dark,—a cloud of appalling blackness overspreading the heavens. Upon this cloud appeared the semblance of a great peacock of fire flying over the town. This suddenly changed to a huge pyramid, that rapidly traversed the sky. Thence arose awful lightnings and thunderings, amid which there fell upon the plain great rocks, some of which weighed 100 lbs. In 1803 a brilliant *fireball* (meteor) was seen traversing Normandy with great velocity, and some moments after, frightful explosions, like the noise of cannon or roll of musketry, were heard coming from a single black cloud hanging in a clear sky; they were prolonged for five or six minutes. These discharges were followed by a great shower of stones, some weighing over 24 lbs. In 1819 a meteor was witnessed in Massachusetts and Maryland, the diameter of which was estimated at half a mile. Its height was thought to be about 25 miles. In July, 1860, a brilliant fireball passed over the state of New York from west to east, and was last seen far out at sea.

SHOOTING STARS.—One of the *earliest accounts* of star-showers is that which relates how, in 472, the sky at Constantinople appeared to be alive with flying stars and meteors. In some Eastern annals we are told that in October, 1202, "the stars appeared like waves upon the sky. They flew about like grasshoppers, and were dispersed from left to right." It is recorded that in the time of King William II. there occurred in England a wonderful shower of stars, which "seemed to fall like rain from heaven. An eye-witness seeing where an aërolite fell, cast water upon it, which was raised in steam with a great noise of boiling." Rastel says concerning it: "By the report of the common people in this kynge's time, diverse great wonders were seene, and therefore the kynge was told by diverse of his familiars, that God was not content with his lyvyng."

In more modern times, the most remarkable accounts are those of the showers of November 12th, 1799, and 1833. Humboldt, in describing the former, says the sky was covered with innumerable fiery trails, which incessantly traversed the sky from north to south. From the beginning of the phenomenon there was not a space in the heavens three times the diameter of the moon which was not filled every instant with the celestial fireworks,—large meteors blending constantly their dazzling brilliancy with the long phosphorescent paths of the shooting stars. *The latter shower* was most brilliant on this continent, and was visible from the lakes to the equator. The scene was one of the most imposing grandeur. Phosphoric lines swept over the sky like the flakes of a sharp snow-storm. Large meteors darted across the heavens, leaving luminous trains behind them that were visible sometimes for half an hour: they generally shed a soft white light; occasionally, however, yellow, green, and other colors varied the scene. Irregular fireballs, almost stationary, glared in the sky; one especially, larger than the moon, hung in mid air over Niagara Falls and mingled its ghastly light with the foam and mist of the cataract. The shower commenced near midnight, but was at its height about 5 A.M. In many sections of the country, the people were terror-stricken by the awful spectacle, and supposed that the end of the world had come.

An inferior shower was seen in 1831 and 1832; and so also in the succeeding years, until 1839. These did not compare in brilliancy with the remarkable phenomenon of 1833.

There was an interval of about 33 or 34 years between the great showers of 1799 and 1833; this seemed to indicate another shower in November, 1866. The people of both hemispheres were literally awake to the subject. Newspapers aroused the most sluggish imagination with thrilling accounts of the scenes presented in 1799 and 1833. Extempore observatories were founded in every convenient point. Watchmen were stationed, and the city bells were to be rung on the appearance of the first wandering celestial visitor. The exact night was not definitely known, but for fear of a mistake, the 11th, 12th, and 13th were generally observed. All painfully testify to those nights being clear and beautiful as moonlight and starlight could make them. The anxious vigils, the fruitless scannings of the sky, the disappointment, the meteors that were dimly *thought* to be seen—all these are recorded in the memory of the temporary astronomers of that year. While, however, the people of America were thus disappointed, there was being enacted in England a display brilliant indeed, though inferior to the one of

1833. The staff at Greenwich Observatory counted about 8,000 meteors; other observers, however, made a much lower estimate. Chambers, in describing the phenomena, says: "Of the large number of descriptions which came under my eye in manuscript and in print, the following is a fair example: 'From 11½ P. M. until 2 A. M. we were much interested in watching the shooting stars; anything so beautiful I never saw, especially about one o'clock, when they were most brilliant;' and so on by the ream." In November, 1867, the long-expected shower was seen in this country, but it failed to satisfy the public expectation. The sky was, however, illumined with shooting stars and meteors, some of which exceeded even Jupiter or Venus in brilliancy.

Number of meteors and shooting stars.—In a paper lately read by Prof. Newton, it is estimated that the average number of meteors that traverse the atmosphere daily, and which are large enough to be visible to the eye on a dark clear night, is 7,500,000; and if to these the telescopic meteors be added, the number would be increased to 400,000,000. In the space traversed by the earth there are, on the average, in each volume the size of our globe (including its atmosphere), as many as 13,000 small bodies, each one capable of furnishing a shooting star visible under favorable circumstances to the naked eye.

Annual periodicity of the star-showers.—On almost any clear night, from five to seven shooting stars may be seen per hour, but in certain months they are much more abundant. Arago names the following principal dates:

April 4–11; 17–25.	October (about) 15.
August 9–11.	November 11–13.

ORIGIN.—Aërolites, meteors, and falling stars all seem to have a common origin. They are produced by small bodies—planets in miniature—which are revolving, like our earth, about the sun. Their orbits intersect the orbit of the earth, and if at any time they reach the point of crossing exactly with the earth, there is a collision. Their mass is so small, that the earth is not jarred any more than is a railway train by a pebble thrown against it.

These small bodies may come near the earth and be drawn to its surface by the power of attraction; or they may simply sweep through the higher regions of the atmosphere, and there escape its grasp; or, finally, they may, under certain conditions, be compelled to revolve many times around the earth as satellites. Indeed, a French astronomer estimates that there is one now circling about the earth at a distance of 5,000 milés. This companion of our moon has a period of three hours and twenty minutes. The average velocity of these meteoric bodies or *bolides*, as they are frequently called, is thirty-six miles per second—much greater than that of Mercury itself. As they sweep through the air, the friction partly arrests their motion, and converts it into heat and light. The body thus becomes visible to us. Its size and direction determine its appearance. If very small, it is consumed in the upper regions, and leaves only the luminous trail of a shooting star. If of large size, it may sweep along at a high elevation, or plunge directly toward the ground. Becoming highly heated in its course, it sheds a vivid light, while, unequally expanding, it explodes, throwing off large fragments which fall to the earth as aërolites, or continue their separate course as meteors. The cinders of the portion consumed rain down on us as fine meteoric dust.

METEORIC RINGS.—These little bodies, it is thought, do not generally revolve individually about the sun, but myriads of them are collected in several rings, and when the earth passes through one of these floating girdles, a star-shower follows. This would account for their regular appearance in certain seasons of the year. In the cut we see how one ring, intersecting the earth's orbit at two points, would account for the August and November showers. Another ring, more inclined to the earth's path, and crossing it nearer the aphelion point, would produce the April showers.

Recent investigators are inclined to the view that there are separate rings for each of the established periods, and that they are very elliptical. The November ring seems to have its perihelion near the ecliptic, and its aphelion beyond the orbit of Uranus; while the August ring extends beyond the solar system. The day of the month in which the great November shower occurs is becoming later at each re-

Fig. 63.

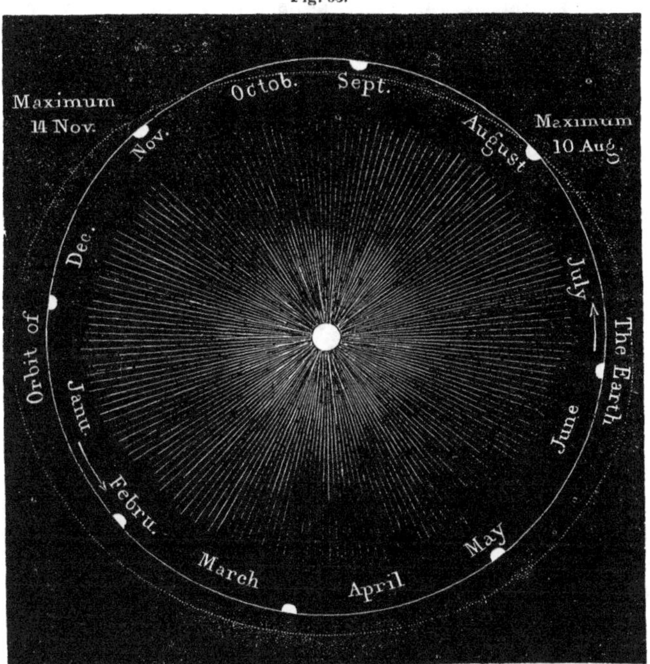

METEORIC RING.

turn; hence it is believed that the nodes of that ring are slowly travelling eastward along the ecliptic. The meteoric bodies are supposed to be quite uniformly distributed through the August stream, but very unequally through the November one. On this account, the former star-showers are quite regular, while the latter vary in brilliancy through periods of $33\frac{1}{4}$ years.

Relation between Meteors and Comets.—The orbit of the November shower is found to be almost identical with that of the comet of 1866; while the August stream is in the track of the comet of 1862. It is a popular theory that these comets are only clusters of meteors crowded so closely together as to be visible by the reflected light of the sun. The single meteors are too small to be seen, except when they plunge into the earth's atmosphere and take fire. On the other hand, Herschel thinks that meteors are the dissipated parts of comets torn into shreds by the sun's attraction.

Radiant Point.—A star (μ) in the blade of the sickle is the point from which the stars in the November shower seem to radiate, while one in Perseus (γ) is the radiant point of the August shower. In the shower of 1866, two observers, who counted the falling stars at the rate of 2,500 per hour, saw only five whose paths, if traced back, would not meet in Leo.

Meteorological effect.—The temperature of August and November is said to be considerably increased by this ring of meteoric bodies, which prevents the heat of the earth from radiating into space. A corresponding decrease of temperature in February and May is caused by the stream or ring of meteors coming between the sun and earth.

HEIGHT.—Herschel estimates the average height of shooting stars above the earth at 73 miles at their appearance and 52 at their disappearance.

WEIGHT.—Prof. Harkness estimates that the average weight of shooting stars does not differ much from *one grain.*

COMETS.

We come now to notice a class of bodies the most fascinating, perhaps, of any in astronomy. The suddenness with which comets flame out in the sky, the enormous dimensions of their fiery trains, the swiftness of their flight, the strange and mysterious forms they assume, their departure as unheralded as their advent—all seem to bid defiance to law, and partake only of the marvellous. Superstitious fears have always been excited by their appearance, and they have been looked upon in every age as

> "Threatening the world with famine, plague, and war;
> To princes, death; to kingdoms, many curses;
> To all estates, inevitable losses;
> To herdsmen, rot; to ploughmen, hapless seasons;
> To sailors, storms; to cities, civil treasons."

Thus the comet of 43 B.C., which appeared just after the assassination of Julius Cæsar, was looked upon by the Romans as a celestial chariot sent to convey his soul heavenward. An old English writer observes: "Cometes signifie corruptions of the ayre. They are signs of earthquakes, of warres, of changyng kyngedomes, great dearthe of corn, yea, a common death of man and beast." Another remarks: "Experience is an eminent evidence that a comet, like a sword, portendeth war; and a hairy comet, or a comet with a beard, denoteth the death of kings, as if God and nature intended by comets to ring the knells of princes, esteeming bells in churches upon earth not sacred enough for such illustrious and eminent performances."

DESCRIPTION.—The term comet signifies a *hairy body*. A comet consists usually of three parts;—the *nucleus*, a bright point in the centre of the *head;* the

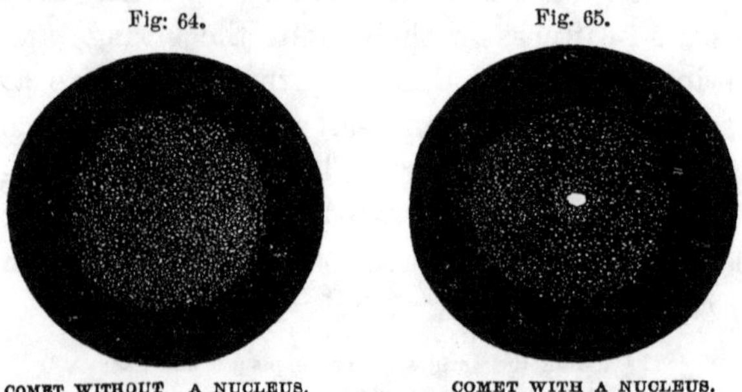

Fig: 64. COMET WITHOUT A NUCLEUS.

Fig. 65. COMET WITH A NUCLEUS.

coma (hair), the cloud-like mass surrounding the nucleus; and the *tail*, a luminous train extending generally in a direction from the sun. There are comets without the tail, and others with several, while some are deprived of even the nucleus. These last consist merely of a fleecy mass, known to be comets from their orbits and rapid motion. Comets are not confined, like the planets, to the limits of the zodiac, but appear in every quarter of the heavens, and move in every conceivable direction. When first seen, the comet resembles a faint spot of light upon the dark background of the sky: as it approaches the sun the brightness increases, and the tail begins to show itself. Generally it is brightest near perihelion, and gradually fades away as it recedes, until it is finally lost, even to the telescope.

THE TIME OF THE GREATEST BRILLIANCY depends somewhat on the position of the earth. If, as represented in the figure, the earth is at *a* when the comet, moving toward perihelion, is at *r*, the comet will appear more distinct than when it is more distant at *s*, although at the latter point it is really brighter. If, however, the earth is at *c* or *b* at the time of perihelion, the comet would be much more conspicuous. Again, if the earth is passing from *a* to *b* during the time the comet is near the sun, it will appear less brilliant than if it were moving from *c* to *d*, as we should then be much nearer it during its greatest illumination.

Fig. 66.

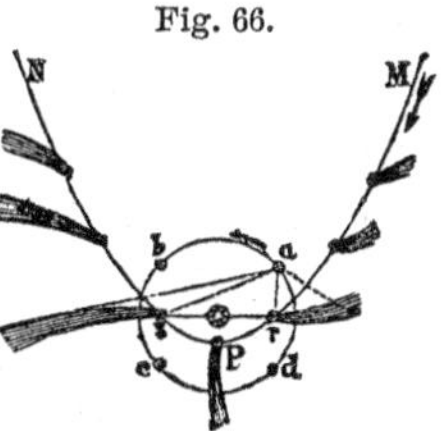

ORBIT OF COMET.

NUMBER OF COMETS.—Kepler remarks that there are as many "comets in the heavens as fish in the sea." Arago has estimated that there are 17,500,000 within the solar system, basing his calculations on the number known to exist between the sun and Mercury. Of this vast number, few are visible to the naked eye, and a still less number attract observation, owing to their inferior size and brilliancy. Many are doubtless lost to our sight by being above the horizon in the daytime. Seneca mentions that during a total solar eclipse, a large and splendid comet suddenly made its appearance near the sun.

Orbits of the Comets.—Comets form a part of the solar system, and are subject to the laws of gravitation. Like the planets, they revolve around the sun, but they differ in the form of their orbits. While the planets move in paths varying but little from circular, and thus never depart so far from the sun as to be invisible to us, the comets travel in extremely elongated (flattened) ellipses, so that they can be observed by us only through a very small portion of their paths. In Fig. 67 are represented the three general classes of their orbits. A comet travelling along an elliptical orbit, though it may pass far from the sun, will yet return within a fixed time; one pursuing either a parabolic or hyperbolic curve cannot return, as the two sides separate from each other more and more. Many of the comets of the first class have been calculated, and they have repeatedly visited our portion of the heavens; while those of the other classes, having once formed part of our system, go away forever, seeking perhaps in the far-off space another sun, which in turn they will abandon as they have our own.

Fig. 67.

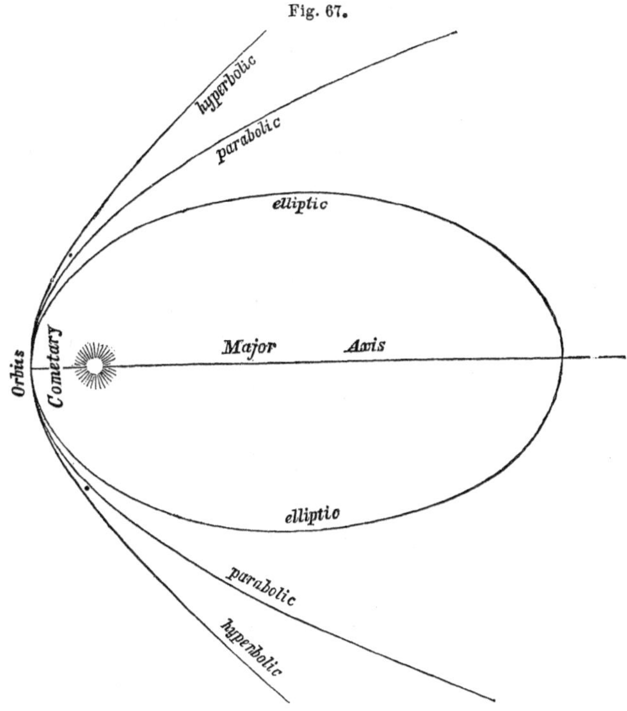

THREE FORMS OF COMETARY ORBITS.

Calculation of a Comet's Return.—As we can observe so small a proportion of the entire orbit, it is very difficult, indeed oftentimes impossible, to decide whether it is an ellipse, hyperbola, or parabola. A few are known to move in clearly elliptical paths, and their movements have been so accurately estimated that it is possible to predict their exact place in the starry vault on any given day and hour. The other comets may never return, or at least not for centuries hence. They may be paying our sun their first visit; or if they have swept through the solar system before, it may have been at so remote a time that no record is preserved, even if it were not before the creation of man. Under these circumstances it is obviously extremely difficult to determine the times of these apparently erratic wanderers; yet, in spite of all these obstacles, some have been tracked far into space beyond the telescopic view. For example, the comet of 1844 is announced to pay a visit to the astronomers of the year of our Lord 101,844. The period of the comet of 1744, is fixed at 122,683 years.

Distance from the Sun.—The comets at their perihelion sweep very near the sun. Thus the one of 1680 came where the temperature was estimated by Newton to be about 2,000 times that of red-hot iron. The nearest approach known is that of the comet of 1843, whose perihelion distance was but about 30,000 miles from the surface of the sun; in fact, it doubled around that body in two hours' time. (Guillemin.) The greatest aphelion distance yet estimated is that of the comet of 1844, which is over 400,000,000,000 miles. The velocity varies, of course, with the position in the orbit. The comet of 1680 moved in perihelion at the rate of over two hundred and seventy-seven miles per second; while in aphelion its velocity is only about six miles per hour.

Density of Comets.—The quantity of matter contained in a comet is exceedingly small. Telescopic stars even are visible through them. The comet of 1770 became entangled among Jupiter's moons, and remained there four months without interfering with their movements in the least; indeed, so far from that, its own orbit was so much changed by the proximity, that from a periodical return of $5\frac{1}{2}$ years, it has not been seen since. The same comet came within 1,400,000 miles of the earth without producing any sensible effect. In 1861, we have good reason to suppose that the earth actually passed through the tail of a comet, its presence being indicated only by a peculiar phosphorescent mist. So that even should our earth run full-tilt against a comet, the shock would be quite imperceptible.* Still, however lightly we may speak of the probability of such a collision, we must remember that there are comets of greater solidity. Donati's, for instance, is estimated to be about $\frac{1}{700}$ the bulk of the earth. The concussion of such a body, moving

* "However dangerous might be the shock of a comet, it might be so slight that it would only do damage at that part of the earth where it actually struck; perhaps even we might cry quits, if, while one kingdom were devastated, the rest of the earth were to enjoy the rarities which a body coming from so far might bring to it. Perhaps we should be very surprised to find that the *débris* of these masses that we despised were formed of gold or diamonds; but who would be the more astonished—we or the comet-dwellers who would be cast on our earth? What strange beings each would find the other!" *Lettre sur la Comète*—(M De Maupertuis.)

with the speed of a cannon-ball, would undoubtedly produce a very sensible effect.

It is not understood whether comets shine by their own or by reflected light. If, however, their nuclei consist of white-hot matter, a passage through such a furnace would be any thing but desirable or satisfactory. After all the calculations of Astronomy, our only safety lies in that Almighty Power which traces the path and guides the course alike of planets and comets: He, whose eye marks the fall of the sparrow, sees as well the flight of the worlds He has created.

Variations in Form and Dimensions.—Comets appear to be subject to constant variations. They are now generally thought to decrease in brilliancy at each successive revolution about the sun. The same comet may present itself sometimes with a tail, and sometimes without. When the comet first appears, there is generally no tail visible, and the light is faint. As it approaches the sun, however, its brightness increases, the tail shoots out from the coma, and grows daily in length and splendor. Supernumerary tails, shorter and less distinct than the principal one, dart out, but they generally soon disappear, as if from lack of material. The tail of the comet of 1843, just after the perihelion, increased in length 5,000,000 miles per day. As the tail thus extended, the nucleus was correspondingly contracted, so that this comet actually "exhausted itself in the manufacture of its own tail."

Remarkable Comets.—Among the many comets celebrated in history, we shall only notice some of those that have appeared in the present century. The *great comet of* 1811 was a magnificent spectacle. The head was 112,000 miles in diameter; the nucleus was 400 miles; while the tail, of a beautiful fan-shape, stretched out 112,000,000 miles. The aphelion distance of this comet is fourteen times that of Neptune, or 40,000,000,000 miles. It is announced to return in thirty centuries! To what profound depths of space, beyond the solar system, beyond the reach of the telescope, must such a journey extend!

The comet of 1835 is commonly known as Halley's comet. This is remarkable as being the first comet whose period of revolution was satisfactorily established. Dr. Halley, on examining the accounts of the great comets of 1531, 1607, and 1682, suspected that they were only the reappearance of the same comet, whose period he fixed at about 75 years. He finally ventured to predict the return of the comet about the end of 1758 or beginning of 1759. Although Halley did not live to see his prophecy fulfilled, great interest was felt in the result. It was not destined, however, for a professional astronomer to be the first to detect the comet. A peasant near Dresden saw it on Christmas night, 1758. The history of this comet, as it has been traced back by its period of seventy-five years, is quite eventful. It was seen in England in 1066, when it was looked upon with dread as the forerunner of the victory of William of Normandy. It was then equal to the full moon in size. In 1456, its tail reached from the horizon to the zenith. It was supposed to indicate the success of Mahomet II., who had already taken Constantinople, and threatened the whole Christian world. Pope Calixtus III., therefore, ordered extra *Ave Marias* to be repeated by everybody, and also the church bells to be rung daily at noon (whence originated the custom now so universal). A prayer was added as follows: "Lord, save us from the devil, the Turk, and the comet." In 1223, it was considered the precursor of the death of Philip Augustus. The first recorded appearance of Halley's comet was B. C. 130, when it was supposed to herald the birth of Mithridates.

The comet of 1843 was so intensely brilliant that it was visible in full daylight. It was so near the sun as "almost to graze his surface."

Encke's comet has a period of only 3⅓ years. A most interesting discovery has been made from observations upon its motion. The comet returns each time to its perihelion about 2½ hours earlier than the most perfect calculations indicate. Hence, Prof. Encke has been led to conjecture that space is filled with a thin, ethereal medium capable of diminishing the centrifugal force, and thus contracting the orbit of a comet.

Donati's comet, which appeared in 1858, was the subject of universal wonder. When first discovered, in June, it was 240,000,000 miles from the earth. In August, traces of a tail were noticed, which expanded in October to about 50,000,000 miles in length. This

Fig. 68.

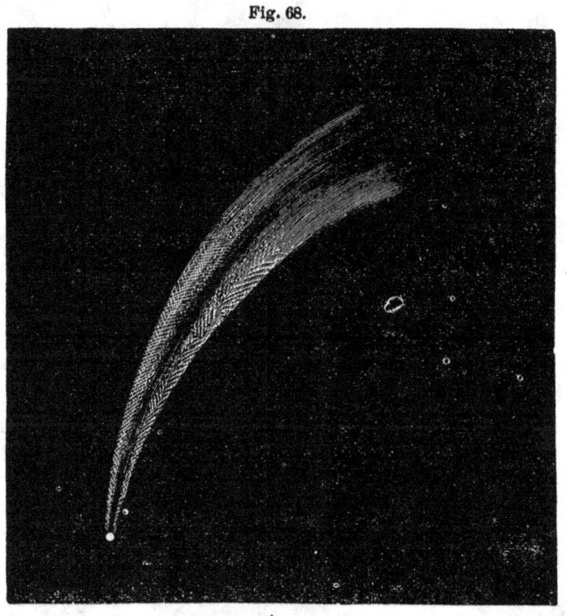

DONATI'S COMET.

comet, though small, has never been exceeded in the brilliancy of the nucleus and the graceful curvature of the tail. It will return in about 2,000 years.

Fig. 69.

ZODIACAL LIGHT.

ZODIACAL LIGHT.

Description.—If we watch the western horizon in March or April, just after sunset, we shall sometimes see the short twilight of that season illuminated by a faint, nebulous light, of a conical shape, flashing upward, often as high as the Pleiades. In September and October, at early dawn, the same appearance can be detected near the eastern horizon. The light can be seen in this latitude only on the most favorable evenings, when the sky is clear and the moon absent. Even then, it will be frequently confounded with the Milky Way or auroral lights. At the base it is of a reddish hue, where it is so bright as very often to efface the smaller stars. In tropical regions the zodiacal light is perpetual, and shines with a brilliancy sufficient, says Humboldt, to cast a sensible glow on the opposite part of the heavens.

ORIGIN.—The commonly received opinion is, that it is caused by a faint cloud-like ring, perhaps a meteoric zone, that surrounds the sun, and only becomes visible to us when the sun himself is hidden below the horizon. Others maintain that, since it has been seen in tropical regions in the east and west simultaneously, it can be explained only on the theory of a "nebulous ring that surrounds the earth within the orbit of the moon."

The Sidereal System.

"He telleth the number of the stars; He calleth them all by their names."

Psalm cxlvii. 4.

THE SIDEREAL SYSTEM.

THE STARS.

In our celestial journey we have reached Neptune, the sentinel outpost of the solar system. We are now 2,750 millions of miles from our sun. Yet we are apparently no nearer the fixed stars than when we first started. They twinkle as serenely there in the far-off sky as to us here on the earth. The heavens by night, with the exception of a few changes in the planets, look perfectly familiar. Between them and us there is a vast chasm which no imagination can bridge; a distance so immense that figures are meaningless, and we can only call it *space*,—so profound that to us it is limitless, though beyond we see other worlds twinkling like distant lights over a waste of waters.

We never see the stars.—This assertion seems almost paradoxical, yet it is strictly true. So far are the stars removed from us, that we see only the light they send, but not the surface of the worlds themselves. They are merely glittering points of light. The most powerful telescope fails to produce a sensible disk. This constitutes a marked point of difference between a planet and a fixed star.

THE ANNUAL PARALLAX OF THE FIXED STARS.—When speaking of this subject on page 139, we said that 183,000,000 miles, or the diameter of the earth's orbit, is taken as the unit for measuring the parallax of the fixed stars. Yet when the stars are viewed from even these extreme points, they manifest so very slight a change of place, that to estimate it is one of the most delicate feats of astronomy.* At the present time, it is considered that the star Alpha (α) Centauri in the southern heavens is the nearest to the earth. Its parallax is judged to be about 1″. Its distance is more than 200,000 times that of the earth from the sun, or *nineteen trillions of miles*. This is probably by no means its extreme distance, but merely the limit *within* which it cannot be, but *beyond* which it must be. These figures convey to our mind no idea of distance. Our imagination fails to grasp the thought, or to picture the vast void across which we are gazing. We remember that light moves at the wonderful rate of 183,000 miles per second. A ray at this speed would plunge out into the abyss beyond Neptune, in one day, six times the distance of that planet

* Prof. Airy says the star which gives the greatest parallax of any, presents the same angle as that at which a circle six-tenths of an inch in diameter would be seen at the distance of a mile!

from the sun. Yet it must sweep on at this prodigious speed, day and night, for three years and nine months to span the gulf and reach a stopping point at the nearest fixed star. "To a spectator standing at *a* Centauri, the entire diameter of the earth's orbit would be hidden by a thread $\frac{1}{25}$ of an inch in diameter, held at a distance of 650 feet from the eye." That is to say, a line 183,000,000 miles long, looked at broadside, would shrink into a mere point. If our sun were removed to that distance, it would shine with a light only equal to that of the north polar star, and would take its place among the constellations as a fixed star.

This, we must remember, is the distance of the *nearest* fixed star. It has been estimated that the average time required for the light of the smallest stars which are visible to the naked eye to reach the earth is about 125 years. What, then, shall we say of those far-distant ones, whose faint light appears as a mere fleecy whiteness even in the most powerful telescopes? The conclusion is irresistible, that the light we receive set out on its sidereal journey far back in the past, perhaps before the creation of man!

Motion of the fixed Stars.—It will aid us still further in comprehending the immense distances of the stars, to learn that though they seem to be fixed, yet they are moving much more swiftly than any of the planets. Thus, Arcturus flies through space at the astonishing rate of about 200,000 miles per hour, or nearly twice that of Mercury, and more than three times that of the earth. Yet, through all our lifetime, we shall never be able to detect any change in its position. It requires three centuries for it to move over the starry vault a space equal to the moon's apparent diameter.

THE STARS ARE SUNS.—The vast distance at which they are known to be, precludes the thought of their shining, like the planets or the moon, by reflecting back the light of our sun. They must be self-luminous, and are doubtless each the centre of a system of planets and satellites.

OUR SUN A STAR.—As we see only the suns of these distant systems, so their inhabitants see only the sun of ours, and that as a *small star*.

OUR SYSTEM ITSELF IN MOTION.—Like all the other stars, our sun is in motion. It is sweeping onward, with its retinue of worlds, 150,000,000 miles per year, toward a point in the constellation Hercules. The Pleiades are thought to be the centre around which this great movement is taking place, but the orbit is so vast and the centre so remote, that nothing definite is yet known.

THE NUMBER OF THE FIXED STARS.—As we look at the heavens on a clear night, the stars seem almost innumerable. To count them, one would think almost as interminable a task as to number the leaves on the trees. It is, therefore, somewhat startling to learn that the entire number visible to the most piercing eyesight, does not exceed 6,000, while few can discern more than 4,000. This illusion may be easily explained, when we remember how the impression of a bright light remains upon the retina, as in the whirling of a firebrand. However, the number

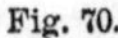

Fig. 70.

A PART OF THE CONSTELLATION OF THE TWINS.

which may be seen with a telescope becomes altogether marvellous. In the cut is shown a portion of the heavens where the naked eye sees but six stars. Could we examine the same region of the sky with more powerful instruments, new constellations would doubtless be descried in the infinite depths of space.

SCINTILLATION.—The twinkling of the fixed stars is due to what is termed in Natural Philosophy "the interference of light." The air being unequally dense, warm, and moist in its various strata, transmits very irregularly the different colors of which white light is composed. Now one color prevails over the rest, and now another, so that the star appears to change color incessantly. As the purity of the air varies, the twinkling of the stars also changes, although it is always greatest near the horizon. Humboldt says that at Cumana, in South America, where the air is remarkably pure and uniform in density, the stars cease to twinkle after they have risen 15° above the horizon. This gives to the celestial vault a peculiarly calm and soft appearance.

MAGNITUDE OF THE STARS.—As the telescope reveals no disk of even the nearest stars, we know nothing of their comparative size. The finest spider's web, placed at the focus of the instrument, hides the star from the eye. When the moon passes in front of a star, the occultation is instantaneous, and not gradual, as in the case of the planets. Classification depends, therefore, upon their relative brightness. The most conspicuous are termed stars of the *first magnitude.* There are about twenty of these. The number of second magnitude stars in the entire heavens is about sixty-five; of the third, about 200; of the fifth, 1,100; and of the sixth,

Fig. 71.

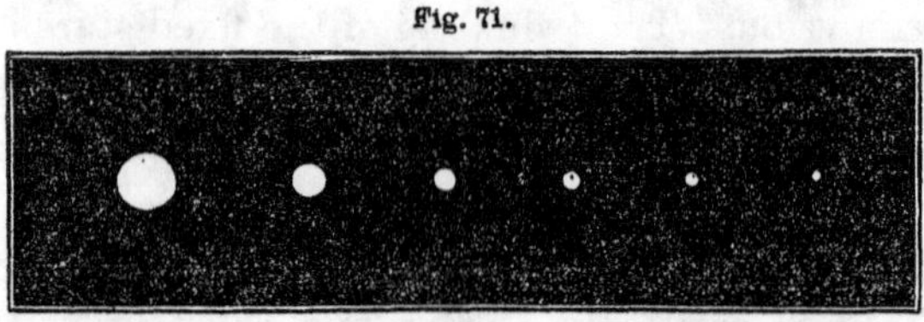

3,200. Few persons can see any smaller stars than those of the fifth or sixth magnitude. The ordinary telescope shows faint stars down to the tenth, while the more powerful instruments reveal those as low as the twentieth magnitude.

THE CAUSE OF THE DIFFERENCE IN THE BRIGHTNESS OF THE STARS.—This may result from a difference in their distance, size, or intrinsic brightness. Whence it follows that the faintest stars may not be the most distant from the earth.

NAMES OF THE STARS.—Many of the brightest stars received proper names at an early date; as Sirius, Arcturus. The stars of each constellation are distinguished by the letters of the Greek alphabet; the brightest being usually called Alpha, the next Beta etc.,—the name of the constellation, in the genitive case, being put after each. Ex., α Arietis, β Lyræ.*

* This means α of Aries, β of Lyra; the genitive case in Latin being equivalent to the preposition *of*.

The Greek Alphabet.

Α	α	Alpha	Ν	ν	Nu
Β	β	Beta	Ξ	ξ	Xi
Γ	γ	Gamma	Ο	ο	Omicron
Δ	δ	Delta	Π	π	Pi
Ε	ε	Epsilon	Ρ	ρ	Rho
Ζ	ζ	Zeta	Σ	ς	Sigma
Η	η	Eta	Τ	τ	Tau
Θ	θ	Theta	Υ	υ	Upsilon
Ι	ι	Iota	Φ	φ	Phi
Κ	κ	Kappa	Χ	χ	Chi
Λ	λ	Lambda	Ψ	ψ	Psi
Μ	μ	Mu	Ω	ω	Omega

When the Greek letters are exhausted, the Roman alphabet is used in the same way. Star catalogues are issued, containing the stars arranged in the order of their Right Ascension, and numbered for convenience of reference. Argelander's Charts have 300,000 stars marked in the northern hemisphere.

The Constellations.—From the earliest ages, the stars have been arranged in constellations, for the purpose of more readily distinguishing them. Some of these groups were named from their supposed resemblance to some figures, such as perching birds, pugnacious bulls, or contorted snakes, while others do honor to the memory of the classic heroes of antiquity.

"Thus monstrous forms, o'er heaven's nocturnal arch,
Seen by the sage, in pomp celestial march;

> See Aries there his glittering bow unfold,
> And raging Taurus toss his horns of gold;
> With bended bow the sullen Archer lowers,
> And there Aquarius comes with all his showers;
> Lions and Centaurs, Gorgons, Hydras rise,
> And gods and heroes blaze along the skies."

With a few exceptions, the likeness is purely fanciful. The heavens are much less of a menagerie than a celestial atlas would make them appear. The division into constellations is a mere relic of barbarism, entirely unworthy of modern civilization. Not only are the figures uncouth, and the origin often frivolous, but the boundaries are not distinct. Stars often occur under different names; while one constellation encroaches upon another. As Chambers well remarks, "Aries should not have a horn in Pisces and a leg in Cetus, nor should 13 Argôs pass through the Unicorn's flank into the Little Dog. 51 Camelopardali might with propriety be extracted from the eye of Auriga, and the ribs of Aquarius released from 46 Capricorni." While, however, the constellations are thus rude and imperfect, there seems little hope of any change. Age gives them a dignity that insures their perpetuation.

Invention of the Constellations.—This goes back into ages of which no record remains. By some it has been ascribed to the Greeks. When the signs of the zodiac were named, they doubtless coincided with the constellations. Aries (the ram) was so called because it rose with the sun in the springtime, and the Chaldean shepherds named it from their flocks, their most valued possession. Then followed in order Taurus (the bull) and Gemini (the twins), called from the herds, which were esteemed next in value. At the summer solstice the sun appears to stop, and, crab-like, to crawl backward; hence the name Cancer (the crab). When the sun is in Leo, the brooks being dry, the lion leaves his lurking-place and becomes a terror to all. Virgo comes next, when the virgins glean in the summer harvest. At the autumnal equinox the days and nights are equally balanced, and this is beautifully represented by Libra (the scales). The vegetation decays in the fall, causing sickness and death; the Scorpion, that stings as it recedes, is suggestive of this Parthian warfare. Sagittarius (the archer) tells of the hunting month. Capricornus (the goat), which delights in climbing lofty precipices, denotes how at the winter solstice the sun begins to climb the sky on his return north. Aquarius (the water-bearer) is a natural emblem of the rainy season. Pisces (the fishes) is the month for fishing.

SIGNS AND CONSTELLATIONS DO NOT AGREE.—By the precession of the equinoxes, as we have before described on page 121, the signs have fallen back along the ecliptic about 30°, so that those stars which were, in the infancy of astronomy, in the sign Aries (♈) are now in Taurus (♉), and those which were in the sign Pisces (♓) are now in Aries (♈).*

* If the teacher put a pin at the centre of Fig. 72, and, drawing a sharp knife between the signs and the constellations, cause the inner part to revolve, the signs may be turned before any constellation, and thus this change be clearly apprehended.

The accompanying cut may illustrate this more clearly.

Fig. 72.

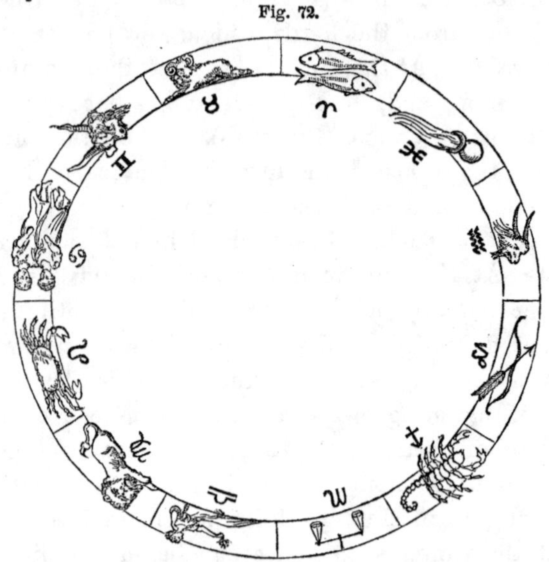

SIGNS AND CONSTELLATIONS, AS THEY NOW COMPARE IN THE HEAVENS, THE FORMER HAVING FALLEN BACK, AND THE LATTER APPARENTLY ADVANCED, 30° EACH.

PERMANENCE OF THE CONSTELLATIONS.—The figures which the stars form, and the general appearance of the constellations, are due to the position we occupy. Could we cross the gulf of space beyond Neptune, the stars now so familiar to us would look strangely enough in their new groupings. As one in riding through a forest sees the trees apparently increase in size and open up to view before him, while they decrease in size and close in behind him, forming clusters and groups which constantly change as he passes along; so, as our earth travels with the solar system on its immense sidereal journey, the stars will grow larger and brighter in front, while those behind us will appear smaller and dimmer. Since, in addition to this, the stars themselves are in motion with varying velocity and in different directions, the constellations must change still more rapidly, so as ultimately to transform entirely the appearance of the heavens. In time, the "bands of Orion" will be loosened, and the "Seven Sisters" will glide apart into remote space. Such are the distances however, that, although these movements have been going on constantly, yet since the creation of man no variation has occurred that is perceptible, save to the watchful astronomer. Nothing in nature is as invariable as the stars. They are the standards of time. Myriads of years must elapse before new star-maps will be required. We need not, then, allow any fear of confusion to disturb us while we study the sky as it is.

Value of the stars in practical life.—"The stars are the landmarks of the universe." They seem to be placed in the heavens by the Creator, not alone to elevate our thoughts and expand our conceptions of the infinite and eternal, but to afford us, amid the constant fluctuations of our own earth, something unchangeable and abiding. Every landmark about us is constantly changing, but over all shine the "eternal stars," each with its place so accurately marked, that to the astronomer and geographer no deception is possible. To the mariner, the heavens become a dial-plate, the figures on its face set with glittering stars, along which the moon travels as a shining hand that marks off the hours with an accuracy no clock can ever rival. Standing on the deck of his vessel, far out at sea, a single observation of the sun or stars decides his location in the waste of waters as accurately as if he were at home, and had caught sight of some old landmark he had known from his boyhood. In all the intricacies of surveying, the stars furnish the only immutable guide. Our clocks vainly strive to keep time with the celestial host. Thus, by a wise provision of Providence, even in the most common affairs of life, are we compelled to look for guidance from the shifting objects of earth up to the heavens above.

THE VIEWS OF THE ANCIENTS.—Standing in the light of our present knowledge, the ideas of the ancients seem almost incredible, and we can hardly understand how they could have been seriously entertained. Anaximenes (550 B. C.) thought that the stars were for ornament, and were nailed like bright studs into the crystalline sphere. Anaxagoras (450 B. C.) considered that they were stones whirled up from the earth by the rapid motion of the ether around us, and that its inflammable properties set them on fire and caused them to shine as stars. Many schools of the Grecian philosophers—the Stoics, Epicureans, etc.—believed that they were celestial fires kept alive by matter that constantly streamed up to them from the centre of the heavens. The stars were at one time said to feed on air; at another, to be the breathing holes of the universe.

THREE ZONES OF STARS.—If we recall what was said on page 104, concerning the paths of the stars and appearance of the heavens at different seasons of the year, we shall see that the constellations are naturally divided into three zones. The *first* embraces those which are visible through the entire year; the *second*, those whose orbits can be seen only in part on any given night; and the *third*, those whose paths just graze our southern horizon, or never pass above it.

THE CONSTELLATIONS.

NORTHERN CIRCUMPOLAR CONSTELLATIONS.—These constellations in our latitude are visible every night. They may be easily traced by holding the book up toward the northern sky in such a way that Polaris and the Dipper on the map and in the heavens agree in position, and then locating the other constellations by comparison. As they revolve about Polaris, their places will vary with every successive night through the year. The cut represents them as they are seen at midnight of the winter solstice. At 6 P. M. of that day the right-hand side of the map should be held downward, and the

Big Dipper will be directly below the north star. At 6 A. M. the left-hand side should be at the bottom, and the Dipper will be above Polaris. From day to day this aspect will change, each star coming

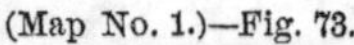
(Map No. 1.)—Fig. 73.

NORTHERN CIRCUMPOLAR CONSTELLATIONS.

a little earlier to the meridian, or to its position on the preceding night. The rate of this progression is six hours, or 90°, in three months.

Ursa Major is represented under the figure of a *great bear.* It contains 138 stars visible to the naked eye. The constellation has been celebrated among all nations. It is remarkable that the shepherds of Chaldea in Asia, and the Iroquois Indians of America, gave to it the same name.

Principal stars.—A noticeable cluster of seven stars—six of the second and one of the fourth magnitude—forms what is familiarly termed "*The Dipper.*" In England it is styled Charles's Wain, from a fancied resemblance to a wagon drawn by three horses tandem. Mizar (ζ) has a minute companion, Alcor, which Humboldt tells us could be rarely seen in Europe. A person with good eyesight may now readily detect it. Megrez (δ), at the junction of the handle and the bowl, is to be marked particularly, since it lies almost exactly in the colure passing through the autumnal equinox. Dubhe and Merak are termed "*The Pointers,*" since they always point out the polar star. The bear's right fore paw and hinder paw are each marked by two small stars, as shown in the cut; a similar pair nearly in line with these denote the left hinder paw (see ξ, Fig. 76). The pairs are 15° apart.

Mythological history.—Diana had a very beautiful attendant named Callisto. Juno, the queen of heaven, becoming jealous of the maid, transformed her into a bear.

"The prostrate wretch lifts up her head in prayer,
Her arms grow shaggy, and deformed with hair;
Her nails are sharpened into pointed claws,
Her hands bear half her weight and turn to paws.
Her lips, that once would tempt a god, begin
To grow distorted in an ugly grin.

And lest the supplicating brute might reach
The ears of Jove, she was deprived of speech.
How did she fear to lodge in woods alone,
And haunt the fields and meadows once her own!
How often would the deep-mouthed dogs pursue,
Whilst from her hounds the frighted hunters flew.

Some time afterward, Callisto's son, Arcas, being out hunting, pursued his mother and was about to transfix her with his uplifted spear, when Jupiter in pity transferred them both to the heavens, and placed them among the constellations as Ursa Major and Ursa Minor.

Ursa Minor is represented under the figure of a small bear. It contains twenty-four stars, of which only three are of the third, and four of the fourth magnitude.

Principal stars.—A cluster of seven stars forms what is termed the "*Little Dipper.*" Three of them are small, and are seen with difficulty. Polaris, at the extremity of the handle, has been known from time immemorial as the North Polar Star. Among the Greeks it was styled Cynosure. Until the mariner's compass came into use, it was the star

"Whose faithful beams conduct the wandering ship
Through the wide desert of the pathless deep."

Polaris does not mark the exact position of the pole, since that is about $1\frac{1}{2}°$ toward the Pointers. This distance will gradually diminish, until in time it will be only $\frac{1}{2}°$: then it will increase again, until in the lapse of ages—12,000 years hence—the brilliant star α Lyræ will fulfil the office of polar star for those who shall then live on the earth.

Curious fact concerning the Pyramids.—Of the nine Pyramids which are standing at Gizeh, Egypt, six have openings facing the north. These lead to straight passages which descend at a uniform angle of about 26° and are parallel with the meridian. If we suppose a person, 4000 years ago, standing at the lower end of one of these passages, and looking out, his eye would strike the sky near the star Thuban, which was then the polar star. The supposed date of the building of these Pyramids (2123 B. C.) agrees with that epoch, and very naturally suggests that the builders had some special design in this peculiar construction.

The distance of Polaris is so great, that though the star is moving through space at the rate of ninety miles per minute, this tremendous speed is imperceptible to us. It requires nearly fifty years for its light to reach the earth; so that when we look at Polaris, we know that the ray which strikes our eye set out on its journey through space half a century ago. We cannot state positively that the star is now in existence, since if it were destroyed to-day it would be fifty years before we should miss it.

Calculation of latitude from Polaris.—By an observer at the equator, Polaris is seen at the horizon. If he advances north, the horizon is depressed and Polaris seems to rise in the heavens. When it has reached the height of a degree, the observer is said to have passed over a degree of latitude on the earth's surface. As he moves further north, the polar star continues to ascend; its distance above the horizon denoting the latitude of each place in succession, until at the north pole, if one could reach that point, Polaris would be seen directly overhead.

Draco is represented under the figure of a long sinuous serpent, stretching between Ursa Major and Ursa Minor, nearly encircling the latter constellation, and finally reaching out its head almost to the body of Hercules.

Principal stars.—Four small stars form a quadrilateral figure at the head; a fifth of the fourth magnitude which is scarcely visible, marks the end of the nose; several scattered groups and delicate triangles of small stars, denote the position of the various coils of the body; thence, an irregular line of stars traces the dragon's tail around between Ursa Major and Ursa Minor. Thuban lying midway between γ of the Little Dipper and ζ of the Big Dipper, is noted as the polar star of forty centuries ago.

Mythological history.—Many accounts are given of the origin of this constellation, as indeed there are of almost every one in the heavens. The prevalent opinion is, that it is the dragon which Cadmus slew. The story is as follows. Jupiter had carried off Európa. Agénor, her father, sent her brother Cadmus in pursuit of his lost sister, bidding him not to return until he was successful in his search. After a time, Cadmus, weary of his wanderings, inquired of the oracle of Apollo concerning the fate of Europa. He was told to cease looking for his sister, to follow a cow as a guide, and when she rested, there to build a city. Hardly had Cadmus stepped out of the temple, when he saw a cow slowly walking along. He followed her until she came upon the broad plains where Thebes afterward stood. Here she stopped. Cadmus wishing to offer a sacrifice to Jupiter in gratitude for the delightful location, sent his servants to seek for water. In a dense grove near by was a fountain guarded by a fierce dragon (DRACO), and sacred to Mars. The Tyrians approaching this and attempting to dip up some water, were attacked, and many of them killed by that enormous serpent, whose head overtopped the tallest trees. Cadmus, becoming impatient, went in search of his men, and on coming to the spring, saw the sad disaster. He forthwith fell upon the monster, and after a severe battle succeeded in slaying him. While standing over his conquered foe, he heard a voice from the ground bidding him take the dragon's teeth and sow them. He obeyed. Scarcely had he finished ere the earth began to move and the points of spears to prick through the surface. Next nodding plumes shook off the clods, and the heads of armed men protruded. Soon a great harvest of warriors covered the entire plain. Cadmus, in terror at the appearance of these giants, whom he termed Sparti (*the Sown*), prepared to attack them, when suddenly they turned upon themselves, and never ceased their warfare until only five of the crowd survived. These making peace with each other, joined Cadmus and assisted him in building the city of Thebes.

Cepheus is represented as a king in regal state, with a crown of stars on his head, while he holds in his hand a sceptre which is extended toward his wife, Cassiopeia. The constellation contains thirty-five stars visible to the naked eye.

Principal stars.—The brightest star is Alderamin (α), in the right shoulder. Alphirk (β), in the girdle, is at the common vertex of several triangles, which point out respectively the left shoulder (ι), the left knee (γ), and the right foot. The head, which lies in the Milky Way, is marked by a delicate little triangle of three stars. This forms, with α, β, and ι, quite a regular quadrilateral figure. A bright little star of the fifth magnitude, close to Polaris, points out the left foot.

Cassiopeia* is represented as a queen seated on her throne. On her right is the king, on her left Perseus, her son-in-law, and above her is Andromeda, her daughter. The constellation contains fifty-five stars visible to the naked eye.

Principal stars.—A line drawn from Megrez (δ), in Ursa Major, through Polaris and continued an equal distance beyond, will strike Caph (β) in Cassiopeia. This star is noticeable as marking, with the others

---

* For mythological history, see Perseus and Andromeda.

named, the equinoctial colure, and as being on the same side of the true pole as Polaris. The principal stars form the figure of an inverted chair, which is very striking and may be easily traced.

EQUATORIAL CONSTELLATIONS.

The constellations we shall now describe lie south of the circumpolar groups. Only a portion of their paths is above our horizon. In using the maps, the observer is supposed to stand with his back toward Polaris, and to be looking directly south. Commencing with the constellation Perseus, so intimately connected with the other members of the royal family just described, we pass eastward in our survey and notice the various constellations as they slowly defile in silent march across the sky. The first map represents the constellations on or near the meridian at nine o'clock in the evening of the winter solstice. On the evening of the autumnal equinox, the left-hand side of the map should be turned downward toward the eastern horizon. On the evening of the vernal equinox, the right-hand side should be turned to the western horizon. At these different times, the stars, though preserving their relative positions, will be diversely inclined to the horizon. As the stars apparently move westward at the rate of 15° per hour, the time of the evening determines what stars will be visible, and also their distances above the horizon.

(Map No. 2)—Fig. 74.

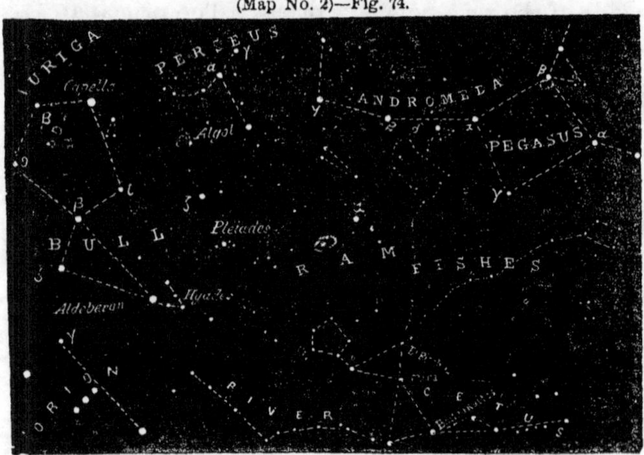

Perseus is represented as brandishing an enormous sword in his right hand, while in his left he holds the head of Medusa. The constellation comprises eighty-one stars visible to the naked eye.

Principal stars.—The most prominent figure is called the *segment of Perseus.* It consists of several stars arranged in a line curving toward Ursa Major. Algenib (α), the brightest of these, is of the second magnitude. Algol, in the midst of a group of small stars, marks the head of Medusa. Between the bright stars of Perseus and Cassiopeia is a beautiful star-cluster visible to the naked eye.

Mythological history.—Perseus, from whom this constellation was named, was the son of Jupiter and Danäe. His grandfather, Acrisius, having been informed by the oracle that his grandson would be the instrument of his death, put the mother and child in a coffer and set them adrift on the sea. Fortunately, they floated near the island Seriphus, where they were rescued and kindly treated by a brother of Polydectes, king of the country. When Perseus had grown up, he was ordered by Polydectes to bring him, as a marriage gift, the head of Medusa. Now Medusa was once a beautiful maiden, who dared to compare her ringlets with those of Minerva; whereupon the goddess changed her locks into hissing serpents, and made her features so hideous, that she turned to stone every living object upon which she fixed her Gorgon gaze. Perseus was at first quite overpowered at the thought of undertaking this enterprise, but was visited by Mercury, who promised to be his guide, and to furnish him with his winged shoes. Minerva loaned him her wonderful shield, that was bright as a mirror. The Nymphs gave him, in addition, Pluto's helmet, which made the bearer invisible. Thus equipped, Perseus mounted into the air and flew to the ocean, where he found the three Gorgons, of whom Medusa was one, asleep. Fearing to gaze in her face, he looked upon the image reflected in Minerva's shield, and with his sword severed Medusa's head from her body. The blood gushed forth, and with it the winged steed PÉGASUS. Grasping the head, Perseus flew away. The other Gorgons awaking, pursued him, but he escaped their search by means of Pluto's helmet. Flying over the wilds of Libya, in his aërial route, drops dripping from the gory head of the monster produced the innumerable serpents for which that country was afterward celebrated.

Andromeda is represented as a beautiful maiden chained to a rock.

Principal stars.—Algenib and Algol in Perseus form, with Almaach (γ) in the left foot of Andromeda, a right-angled triangle opening toward Cassiopeia. This figure is so perfect, that the stars may be easily recognized. The girdle is pointed out by Merach (β), and two other stars which form a line slightly curving toward the right foot. The breast is denoted by a very delicate triangle composed of three stars, δ of the fourth magnitude, another of the fifth magnitude just south, and an exceedingly minute star a little at the west. Alpheratz (α), in the head of Andromeda, belongs also to PEGASUS. This star, with three others, all of the second magnitude, constitute the "*Great Square of Pégasus.*" Their names are Algenib (γ), Markab (α), and Scheat (β). The brightest stars of these two constellations form a figure strikingly like the Big Dipper. Algenib and Alpheratz lie in the equinoctial colure which passes through Caph.

Mythological history.—Cassiopeia had boasted that her daughter Andromeda was fairer than the Sea-nymphs. They appealed, in great indignation, to Neptune, who sent a sea-monster (CETUS) to devastate the coast of Ethiopia. To appease the deities, her father Cepheus was directed by the oracle to bind his daughter to a rock, to be devoured by Cetus. Perseus returning from the destruction of Medusa, saw Andromeda in her forlorn condition. Struck by her beauty and tears, he offered to liberate her at the price of her hand. Her parents consented joyfully, and, in addition, offered a royal dower. Perseus slew the terrible monster, and freeing Andromeda, restored her to her parents. All the prominent actors in this scene were honored with seats among the constellations. The Sea-nymphs, it is said, in petty spite of Cassiopeia, prevailed that she should be placed where for half of the time she hangs with her head downward—a fit lesson of humility. Cepheus, her husband, shares in her punishment.

Aries, the *ram*, was anciently the first constellation of the zodiac. It is now the *first sign*, but the *second constellation.* On account of the precession of the equinoxes, the constellation Pisces occupies the first sign.

Principal stars.—The most noted star is α Arietis (Alpha of Aries, more commonly called simply Arietis), in the right horn. This lies near the path of the moon and is one of the stars from which longitude is reckoned. A line drawn from Almaach to Arietis will pass through a beautiful figure of three stars called the THE TRIANGLES.

Mythological history.—Phryxus and Helle were the children of Athamas, king of Thessaly. Being persecuted by Ino, their step-mother, they were compelled to flee for safety. Mercury provided them a ram which bore a golden fleece. The children were no sooner placed on his back than he vaulted into the heavens. In their aërial journey Helle becoming dizzy fell off into the sea, which was afterward called the Hellespont, now the Dardanelles. Phryxus coming in safety to Colchis, on the eastern shore of the Black Sea, offered the ram in sacrifice to Jupiter, and gave the golden fleece to Aetes, his protector. The Argonautic expedition in pursuit of this golden fleece, by Jason and his followers, is one of the most romantic of mythological stories. It is, undoubtedly, a fanciful account of the first important maritime expedition. Rich spoils were the prizes to be secured.

Taurus consists only of the head and shoulders of a *bull*, which is represented in the act of plunging at Orion.

Principal stars.—The *Hyades*, a beautiful cluster in the head, forms a distinct V. The brightest of these is Aldebaran, a fiery red star of the first magnitude. The *Pleiades*,* or the "Seven Sisters," as it is sometimes termed, is the most conspicuous group in the heavens. It contains a large number of stars, six of which are visible to the naked eye. There were said to have been anciently seven, but Electra left her place that she might not behold the ruin of Troy, which was founded by her son Dar-

* Job, xxxviii. 31; Amos, v. 8.

danus. Others say that the "*lost Pleiad*" was Merope, who married a mortal. Alcyone is the most distinctly seen. El Nath (β) and ζ point out the horns of Taurus.

Mythological history.—This is the animal whose form Jupiter assumed when he bore off Europa. The Pleiades were the daughters of Atlas, and Nymphs of Diana's train. They were distinguished for their unblemished virtue and mutual affection. The hunter ORÍON having pursued them one day, they prayed to the gods in their distress. Jupiter in pity transferred them to the heavens.

Auriga, the *Charioteer* or *Wagoner,* is represented as a man resting one foot on a horn of Taurus, and holding a goat and kids in his left hand and a bridle in his right.

The principal stars are arranged in an irregular five-sided figure, which is very noticeable. Capella, the goat-star, is of the first magnitude. It travels in its orbit 1,800 miles per minute; and it takes seventy-two years, or a man's lifetime, for its light to reach the earth. Near by is a delicate triangle formed of three small stars, called *the Kids.* Menkalini (β) is in the right shoulder, θ in the right hand, β (common to Auriga and Taurus) the right foot and ι the left foot. Capella, β, and δ (a star in the head) form a triangle. The origin of this constellation is unknown.

Pisces, the *fishes,* is represented by two fishes tied together by a long ribbon. It consists of small stars, which can be traced only upon a clear night, and in the absence of the moon.

Cetus, the *whale*, is a huge sea-monster, slowly ploughing his way westward, midway between the horizon and the zenith. It may easily be found, on a clear night, by means of the numerous figures given in the map.

(Map No. 3)—Fig. 75.

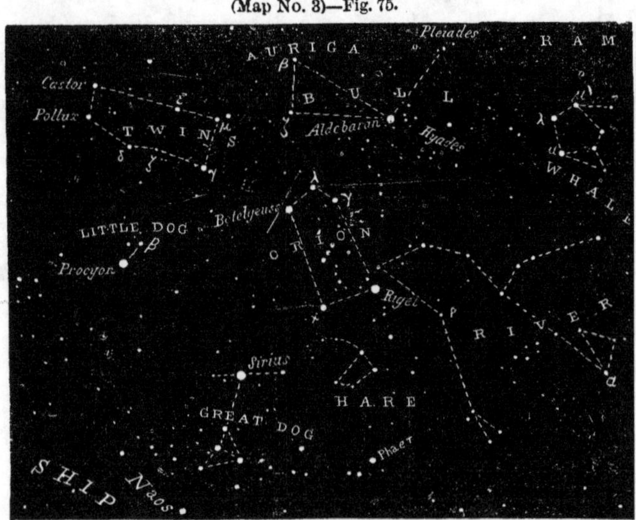

Gemini, the *Twins*, represents the twin brothers Castor and Pollux.

The principal stars are Castor and Pollux, which are of the first and second magnitudes. The latter is also one of the stars from which longitude is reckoned by means of the Nautical Almanac. The constellation is clearly distinguished by means of two nearly parallel rows of stars, which by a slight effort of the imagination may be extended into the constellations Taurus and Orion.

Mythological history.—Castor and Pollux were noted—the former for his skill in training horses, the latter for boxing. They were tenderly attached to each other, and were inseparable in all their adventures. They accompanied Jason on the Argonautic expedition. A storm having arisen on this voyage, Orpheus played on his wonderful lyre and prayed to the gods; whereupon the tempest was stilled, and star-like flames shone upon the heads of the twin-brothers. Sailors, therefore, considered them as patron deities,* and the balls of electric flame seen on masts and shrouds, now called St. Elmo's fire, were named after them. Afterward, Castor was slain. Pollux being inconsolable, Jupiter offered to take him up to Olympus, or to let him share his immortality with his brother. Pollux preferred the latter, and so the brothers pass alternately one day under the earth, and the next in the Elysian Fields. Not only did sailors thus think them to watch over navigation, but they were believed to return, mounted on snow-white steeds and clad in rare armor, to take part in the hard-fought battle-fields of the Romans.

> "Back comes the chief in triumph,
> Who in the hour of fight
> Hath seen the great Twin Brethren,
> In harness on his right.

---

* Acts, xxviii. 11.

> Safe comes the ship to haven,
> Through billows and through gales,
> If once the great Twin Brethren
> Sit shining on the sails."—*Lays of Ancient Rome.*

Orion is represented under the figure of a hunter assaulting Taurus. He has a sword in his belt, a club in his right hand, and the skin of a lion in his left. This is one of the most clearly defined and conspicuous constellations in the heavens.

Principal stars.—Four brilliant stars, in the form of a parallelogram, mark the outlines of Orion. Betelgeuse, a beautiful ruddy star of the first magnitude, is in the right shoulder; Bellatrix (γ), of the second magnitude, is in the left shoulder; Rigel, of the first magnitude, is in the left foot; and Saiph (κ), of the third magnitude, is in the right knee. Two small stars near λ form with it a small triangle, which is itself the vertex of a larger triangle composed of λ, γ, and Betelgeuse. Near the centre of the parallelogram are three stars forming "*the Belt of Orion*," called also the "Bands of Orion" (Job, xxxviii. 31), Jacob's rod, but more commonly the "Ell and Yard." They received the last name because they form a line just 3° long, divided in equal parts by a star in the centre. These divisions are useful for measuring the distances of the stars. Running from the belt southward, is an irregular line of stars which marks the sword; and west of Bellatrix is a curved line denoting the lion's skin. South of Orion are four stars forming a beautiful figure styled the HARE.

Mythological History.—Orion was a famous hunter. Becoming enamored of Merope, he desired to marry her. Œnopion, her father, opposing the choice, took a favorable opportunity and put out the eyes of the unwelcome suitor. The blinded hero followed the sound of a Cyclop's hammer until he came to Vulcan's forge. He, taking pity, instructed Kedalion to conduct him to the abode of the sun. Placing his guide on his shoulder, Orion proceeded to the east, and at a favorable place

> "Climbing up a narrow gorge,
> Fixed his blank eyes upon the sun."

The healing beams restored him to sight. As a punishment for having profanely boasted that he was able to conquer any animal the earth could produce, he was bitten in the heel by a scorpion. Afterward, Diana placed him among the stars; where SIRIUS and PROCYON, his dogs, follow him, the PLEIADES fly before him, and far remote is the SCORPION, by whose bite he perished.

Canis Major and ***Canis Minor*** contain each a single star of the first magnitude, Sirius and Prócyon. These two, with Betelgeuse, Phaet in the Dove, and Naos in the Ship, form a huge figure known as the Egyptian X. Sirius, the dog-star, is the most brilliant star in the heavens. It travels at the rate of 840 miles per minute. Twenty-two years are required for its light to reach the earth; its distance being estimated at 1,375,000 times that of the sun from us. If its intrinsic brilliancy be the same as that of our sun, its diameter at that distance must be fifteen times as great, or 12,000,000 miles. Probably these estimates fall far below the reality of this magnificent orb.

(Map No. 4)– Fig. 76.

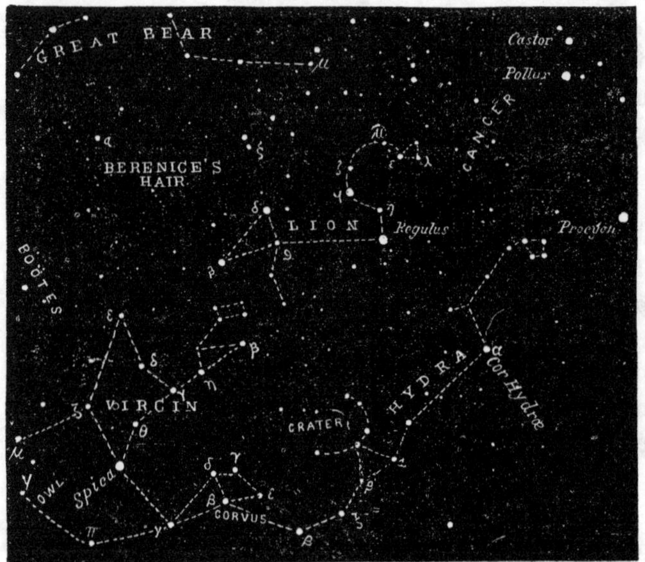

Leo is represented as a rampant lion. It is one of the most beautiful constellations in the zodiac.

The principal stars are arranged in the form of a sickle. Regulus, in the handle, is a brilliant star of the first magnitude. It is one of the stars from which longitude is reckoned. It is almost exactly in the ecliptic. Zosma (δ) lies in the back of the lion, θ in the thigh, and Denebola, a star of the second magnitude, in the brush of the tail.

Cancer includes the stars which lie irregularly scattered between Gemini, Head of Hydra, Procyon, and Leo. In the midst of these is a luminous spot, called Presepe or the Beehive, which an ordinary glass will resolve into stars.

Virgo is represented as a beautiful maiden with folded wings, bearing in her left hand an ear of corn.

The principal star is Spica Virginis, in the ear of corn. It is of the first magnitude, and is used for determining longitude at sea. Denebola, Cor Caroli, (α), Arcturus (Map No. 5), and Spica form a figure about 50° in length from north to south, called the Diamond of Virgo. The other stars may be easily traced by means of the map.

Mythological History.—Virgo was the goddess Astræa. According to the poets, the early history of man was the golden age. It was a time of innocence and truth. The gods dwelt among men, and perpetual Spring delighted the earth. Next came the silver age, less tranquil and serene, but still the gods lingered and happiness prevailed. Then followed the brazen and iron ages, when wickedness reigned supreme. The earth was wet with slaughter The gods left the abodes of men one by one, Astræa alone remaining; until finally she too, last of all the immortals, bade the earth farewell. Jupiter thereupon placed her among the constellations.

Hydra is a long straggling serpent having its head near Procyon and extending its tail beyond Virgo, a total distance of more than 100°.

The principal star is Cor Hydræ, of the second magnitude. It is a lone star, and may be easily found by a line drawn from γ Leonis through Regulus, and continued about 23°. The head is marked by a rhomboidal figure of four stars of the fourth magnitude lying near Procyon. Several delicate triangles may be formed of them and other small stars lying near. The Crater or Cup is a beautiful and very striking semicircle of six stars of the fourth magnitude directly south of θ Leonis. Corvus, the raven, lies 15° east of the Cup. ε Corvi is in the equinoctial colure.

Mythological history.—Hydra was a fearful serpent which in ancient times infested the lake Lerna. Its destruction constituted one of the twelve labors of Hercules. The Crow was formerly white, it is said, but was changed to its raven tint on account of its proneness to tale-bearing.

Cor Caroli (α) is marked by a line passing from Benetnasch (η) through Berenice's Hair to Denebola (β).

Berenice's Hair is a beautiful cluster midway between Cor Caroli and Denebola. Near by is a single bright star of the fourth magnitude.

Mythological history.—Berenice was the wife of Ptolemy. Her husband going upon a dangerous expedition, she promised to consecrate her beautiful tresses to Venus if he should return in safety. Soon after the fulfilment of this vow the hair disappeared from the temple where it had been deposited Berenice being much disquieted at this loss, Conon, the astronomer, announced that the locks had been transferred to the heavens. In proof of which, he pointed out this cluster of hitherto unnamed stars. All parties were satisfied with this happy termination of the difficulty.

(Map No. 5)—Fig. 77.

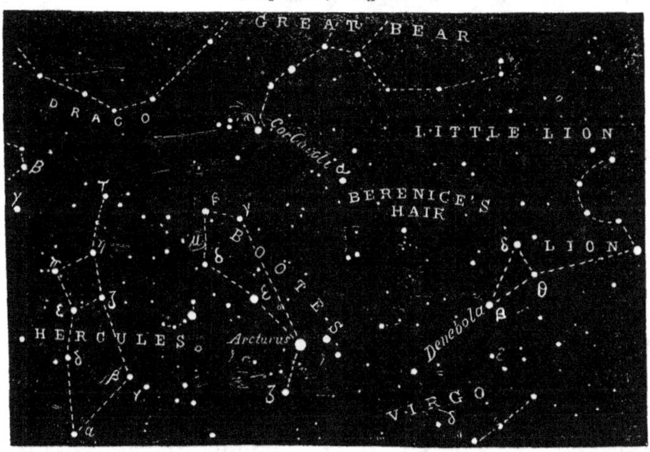

Boötes, the bear-driver, is represented as a huntsman grasping a club in his right hand, while in his left he holds by the leash his two greyhounds, with which he is pursuing the Great Bear continually around the north pole.

Principal stars.—Arcturus,* a magnificent star

---

* Job, ix. 9.

of the first magnitude, is in the left knee. It forms a triangle with Denebola and Spica, and also one with Denebola and Cor Caroli. It travels in its orbit fifty-four miles per second, or more than three times as fast as the earth. Its light reaches the earth in about twenty-six years. Mirac (ϵ) lies in the girdle, δ in the right shoulder, Alkaturops (μ) in the club, β in the head, and Seginus (γ) in the left shoulder. Seginus forms with Cor Caroli and Arcturus a triangle, right-angled at Seginus. Three small stars in the left hand of Boötes lie near Benetnasch.

Mythological history.—Boötes is supposed to have been Arcas, the son of Callisto. (See Ursa Major.)

Hercules is represented as a warrior clad in the skin of the Nemæan lion, holding a club in his right hand and the dog Cerberus in his left. His foot is near the head of Draco, while his head lies 38° south, and his club reaches 10 degrees beyond.

The principal star is Ras Algethi (α of Hercules and ζ of Serpentarius). This forms a triangle with β and δ. A peculiar figure of four stars (π, η, ζ, ϵ), north of these, marks the body. (See Maps, Nos. 5, 7, and 7.) The left knee is pointed out by θ, and the left foot by *y*.

Mythological history.—This constellation immortalizes the name of one of the greatest heroes of antiquity. Hercules was the son of Jupiter and Alcmena. While he was yet lying in his cradle, Juno, in her jealousy, sent two serpents to destroy him. The precocious infant, however, strangled them with his hands. By the cunning artifice of Juno, Hercules was made subject to Eurystheus, his elder half-brother, and compelled to perform all his commands. Eurystheus enjoined upon him a series of the most difficult and dangerous enterprises which could be conceived. These are termed the "Twelve Labors of Hercules." Having completed these tasks, he afterward achieved others equally celebrated. Near the close of his life he killed the centaur Nessus. The dying monster charged Dejanira, the wife of Hercules, to preserve a portion of his blood as a charm to use in case the love of her husband should ever fail her. In time, Dejanira thought she needed the potion, and Hercules having sent for a white robe to wear at a sacrifice, she steeped the garment in the blood of Nessus. No sooner had Hercules put on the fatal robe than the venom stung his bones and boiled through his veins. He attempted to tear it off, but in vain. It stuck to his flesh, and tore off great pieces of his body. The hero finding he must die, ascended Mount Œta, where he erected a funeral pyre, spread out the skin of the Nemæan lion, and laid himself down upon it. Philoctetes applied the torch. With perfect serenity of countenance Hercules awaited approaching death—

"Till the god, the earthly part forsaken,
From the man in flames asunder taken,
Drank the heavenly ether's purer breath.
Joyous in the new unwonted lightness
Soared he upward to celestial brightness,
Earth's dark, heavy burden lost in death."

SCHILLER.

(Map No. 6)—Fig. 78.

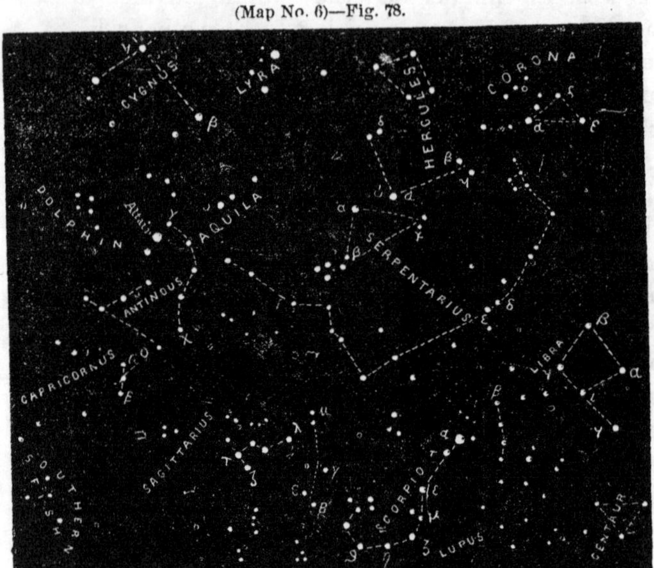

Corona consists of six stars arranged in a semicircular form. The brightest of these is Alphacca. This makes a triangle, with Mirac (ε) and δ in Boötes. It forms a similar figure with Mirac and Arcturus.

Serpentarius, or ***Ophiuchus,*** the serpent-bearer, is represented under the figure of a man grasping in both hands a prodigious serpent, which is writhing in his grasp.

Principal stars.—Ras Alhague (α), in the head, is of the second magnitude. It is about 5° from Ras Algethi. They form a pair of stars conspicuous like the pairs in Gemini, Canis Minor, Canis Major, etc. β marks the right shoulder, and χ the left. There is a small cluster near β, called TAURUS PONIATOWSKII. An irregular square of four stars, near γ Herculis, denotes the head of the serpent.

Mythological history.—This constellation perpetuates the memory of Æsculapius, the father of medicine. He was so skilful that he restored several to life; whereupon Pluto complained to Jupiter that his kingdom was in danger of being depopulated. Therefore Jupiter struck him with a thunderbolt, but afterward placed him among the constellations. Serpents were sacred to Æsculapius, because of the superstitious idea that they have the power of renewing their youth by changing their skin.

Libra represents the scales of Astræa (Virgo), the goddess of justice. It may be recognized by the quadrilateral figure formed by its four principal stars.

Scorpio is represented under the figure of a huge scorpion, stretching through 25°. It is a most interesting constellation.

Principal stars.—Antares (α) is a fiery red star of the first magnitude. It marks the heart of the Scorpion. The head is indicated by several stars, the most prominent of which is β, arranged in a line slightly curved. The tail may be easily traced by a series of stars which wind around through the Milky Way in a very beautiful manner.

Mythological history.—This is the scorpion that sprang out of the earth at the command of Juno, and stung Orion. Scorpio and Orion are so placed among the constellations that they never appear in the heavens together.

Sagittarius, the archer, is represented as a centaur with his bow bent, as if about to let fly an arrow at Scorpio.

Principal stars.—A row of stars from μ to β marks the bow: another from γ eastward points out the arrow and the right arm drawn back in bending the bow. North of τ, two stars of the fourth magnitude denote the head of the centaur. The "*Milk Dipper*," so called because the handle lies in the Milky Way, is a very striking figure.

Mythological history.—This constellation is named in honor of Chiron, one of the centaurs. These monsters were represented as men from the head to the loins, while the remainder of the body was that of a horse—of which animal the ancients had so high an opinion that this union was not considered in the least degrading. Chiron was renowned for his skill in music, medicine, hunting, and the art of prophecy. The most distinguished heroes of mythology were among his pupils. He taught Æsculapius physic, Apollo music, and Hercules astronomy. At his death, the centaur furnished Dejanira with the information which proved so fatal to Hercules.

Capricornus contains no very conspicuous stars. The SOUTHERN FISH (No. 6) has one star of the first magnitude, Fomalhaut (α, No. 7), which on a clear summer evening may be seen in the south midway to the zenith. ANTINOUS AND THE EAGLE is a double constellation. It contains a beautiful star of the first magnitude, Altair. This is conspicuous, as being the centre one in a row of three bright stars. A similar row, the first star of which is named ζ, denotes the tail of the eagle, the last star lying in Cerberus. The DOLPHIN is a beautiful little cluster in the form of a diamond. It is sometimes called "*Job's Coffin.*"

(Map No. 7)—Fig. 79.

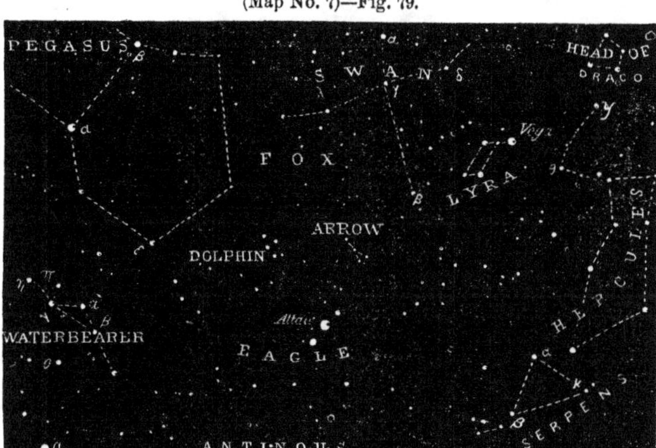

Cygnus, the swan, is a remarkable group of stars, the principal ones being so arranged as to form a large and beautiful cross. The upright piece lies along the Milky Way. It is composed of four stars, three of which, Deneb (α), γ, and β, are bright, while the fourth is a variable star. In this constellation, No. 61, a minute star, scarcely visible to the naked eye, is noted as being the nearest to the earth of any of the fixed stars in the northern hemisphere.

Lyra, the harp, contains one brilliant blue star, Vega. Close by it is a parallelogram of four smaller stars, by which it may be easily recognized. This is the celestial lyre upon which Orpheus discoursed such ravishing music that wild beasts forgot their fierceness and gathered about him to listen, while the rivers ceased to flow, and the very rocks and trees stood entranced.

THE SOUTHERN CONSTELLATIONS

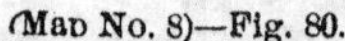
(Map No. 8)—Fig. 80.

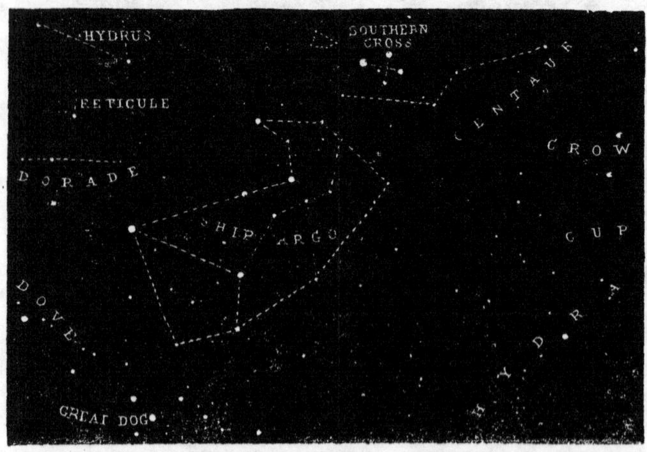

We now imagine ourselves viewing the stars visible only to a person south of the equator. The constellations are reversed with reference to the horizon. The two stars which, in the northern hemisphere, compose the base of the parallelogram in Orion, form here the upper side. Sirius is above Orion. All the northern circumpolar constellations are hidden from view. At the southern pole there is no conspicuous star, but the richness and number of the neighboring stars compensate this deficiency, and give to the heavens an incomparable splendor. Here is the magnificent constellation Argo, in which we find Canopus, looked upon in ancient times as

(Map No. 9)—Fig. 81.

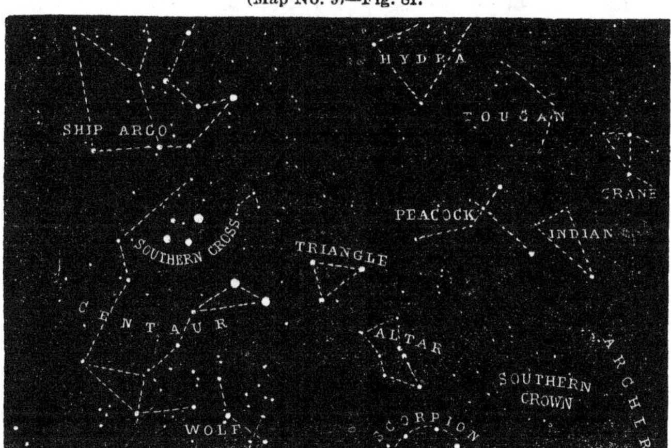

next to Sirius in brilliancy: η, a variable star, now surpasses it in brightness.

Nearly at the height of the south pole blazes the SOUTHERN CROSS; below is the CENTAUR, containing two stars of the first magnitude and five of the second; and above is Hydrus, where shines Achernar, another beautiful star of the first magnitude.

DOUBLE STARS, COLORED STARS, NEBULÆ, ETC.

DOUBLE STARS.—To the naked eye all the stars appear single. With the telescope, over 6,000 have been found to be double. Thus, Polaris consists of two stars about 18″ apart, Rigel has a companion about 10″ from it, and Sirius one distant 7″. A good opera-glass will separate ε Lyræ into two components. In case two stars happen to lie in the same straight line from us, though at immense distances from each other, their light will blend. They will be seen by the naked eye as a single star, and by the telescope as a double star. They are called *optical double* stars. Over 650, however, of the double stars have been found to be *physically* connected. Each double star of this class forms a binary system of two suns revolving in an elliptical orbit about their common centre of gravity, like the planets in the solar system, in accordance with Newton's law of gravitation. In a few instances there are combinations of *triple, quadruple*, and even *septuple* stars. Thus ε Lyræ is a *double-double* star, while θ Orionis is a system of seven suns. The components of a double star commonly differ in brightness; so that frequently the fainter one is nearly lost in the brilliancy of its companion sun.

The periods of some of these systems have been ascertained. Thus, ξ Ursæ is a double star, and the two stars of which it is composed have performed an entire revolution about each other since they were known to be connected. There are only eight binary stars whose periods are less than a century, while 325 have periods which seem to extend one thousand years.

Orbits.—It is not possible to estimate the dimensions of the orbits of the double stars, until their distances from us are known. Taking the estimated distance of 61 Cygni (550,000 times the sun's mean distance from the earth) as a basis, the companions of that system cannot cultivate a very intimate acquaintance, since they must be over a billion miles apart. From these data, astronomers have even attempted to calculate the *mass* of some of the double stars. 61 Cygni, although scarcely visible to the naked eye, and known to be the second nearest to us of any of the fixed stars, is yet estimated to weigh one-third as much as our sun.

COLORED STARS.—We have already noticed that the stars are of various colors. Sirius is white, Antares red, and Capella yellow; while Lyra has a blue tint, and Castor a green one. In the pure transparent atmosphere of tropical regions, the colors are far more brilliant. There, oftentimes, the nocturnal sky is a blaze of jewels,—the stars glittering with the green of the emerald, the blue of the amethyst, and the red of the topaz. In our latitudes, there are no stars visible to the naked eye which are decidedly blue or green. In the double and multiple stars, every color is presented in all its richness and beauty. We find also combinations of colors complementary to each other. Here is a green star with a blood-red companion: here an orange and blue sun- there a yellow and purple one. The triple star γ Andromedæ, is formed of an orange-red sun and two others of an emerald green. Every tint that blooms in the flowers of summer, flames out in the stars at night. "The rainbow flowers of the footstool and the starry flowers of the throne," proclaim their common Author; while rainbow, flower, and star alike evince the same Divine love of the beautiful.

As to the effects produced in a system having colored suns we can hardly conceive. Take a planet revolving about ψ Cassiopeiæ for instance. This is illuminated by a red, a blue, and a green sun. Sometimes, by the succession of these suns, a cheerful green day would present a charming relief to a fiery red one; and that might be still further subdued by a gentle blue one. The odd contrasts of color and the vicissitudes of extreme heat and cold which obtain on such a world, present a picture which our fancy can sketch better than words can paint. The colors of the stars change. Sirius was anciently red. It is now unmistakably white. There are two double stars which were described by Herschel as white; they are each now composed of a golden-yellow and a greenish star.

VARIABLE STARS.—These are stars which have periodic changes of brilliancy. There are many of this class, of which the following are most conspicuous. ALGOL, in the head of Medusa, is a star of the second magnitude for about two and a half days, when it suddenly decreases, and in three and a half hours descends to the fourth magnitude. It then rekindles, and in three and a half hours again is as brilliant as ever. MIRA, the *wonderful*, a star in the Whale, has a period of eleven months. Its irregularities are very curious and fickle. It is ordinarily of the second magnitude for about fifteen days. It then decreases for three months, until it is reduced to the 9th magnitude. This period of darkness lasts five months; it then rebrightens for three months, until it regains its former lustre. Occasionally, however, it fails to brighten at all beyond the fourth magnitude, while on one occasion its light was almost equal to that of Aldebaran. Sometimes no perceptible change takes place for a month; then again, there is a sensible alteration in a few days.

The reason of this variability is not understood. It has been suggested, in the case of Mira, that it may be a globe revolving on its axis, and that different portions of its surface, illuminated to different degrees of intensity, are thus presented to us. Others have conceived that there may be satellites revolving about these suns, and that when their dark bodies interpose between the stars and our earth, they eclipse their light wholly or in part.

TEMPORARY STARS.—These are stars which suddenly blaze out in the heavens, and then gradually fade away. The most celebrated one of this class burst forth in Cassiopeia, in the year 1572. Tycho Brahé says: "One night as I was examining the celestial vault, I saw with unspeakable astonishment a star of extraordinary brightness in Cassiopeia Struck with surprise, I could scarcely believe my eyes. To convince myself that there was no illusion, I called the workmen of my laboratory and the passers-by, and asked them if they saw the star which had so suddenly made its appearance." It was more brilliant than Sirius or Jupiter even, and could be compared only with Venus at her quadrature, except that it twinkled wonderfully. It was seen distinctly at midday. Its color was at first white, then yellow, and finally red. Its brightness decreased gradually until the spring of 1574, when the star disappeared from view and has not since been seen. As two brilliant stars had previously appeared in Cassiopeia, at intervals of about three centuries, they have been thought, by some, to be identical, and that it is only a variable star of long period.

Since the discovery of Tycho Brahé, numerous instances are recorded of stars which have suddenly burst forth, and then either faded out entirely, or remained only as faint telescopic objects. In the latter case they are termed *new stars.* One of this kind appeared in Corona Borealis, in 1866. At first it was of the second magnitude, but in a week changed to the fourth, and in a month diminished to the 9th. Strangely, too, some stars have disappeared from the heavens, and are styled *lost stars.* These changes which are thus constantly taking place are calculated to make the term "eternal stars" seem a very indefinite phrase.

Explanation.—These phenomena are as yet little understood. A revolution about the axis would fail to explain the changes in color, besides being in itself a very unaccountable supposition. Some think that these stars revolve in enormous orbits of such eccentricity that at their most distant points they fade out of sight. Arago has shown, in reply to this, that for a star to decrease in brightness from the first magnitude to the second by simply moving directly from us, would require six years, even if it should speed away with the velocity of light. As we have just seen, the star of 1866 underwent this change in brilliancy in a week.

The mind cannot help wondering if they are not instances of enormous conflagrations in which a world is overwhelmed in ruin! The investigations of spectrum analysis indicate that the star of 1866 consisted of *burning hydrogen gas.* We can suppose that this was evolved by some convulsion, and taking fire, wrapped in flames the entire globe. This need not involve the idea of destruction, but only a change of form. In this manner a dark star may become luminous, or a bright one may be extinguished.

Thus do we see that the process of apparent creation and destruction is going on in the heavens immediately before the eye of the astronomer. New stars flash into light, old stars are lost, worlds burst into flame, and their glowing embers fade into darkness. Are they re-created into new worlds? We know not. We only perceive that the same Almighty power which fitted up this earth for our home is yet at work among the worlds about us, and we are thus witnesses of His eternal presence.

STAR CLUSTERS.—These are groups of stars so massed together as to present a hazy, cloud-like appearance. Several of them have been already named—the Pleiades, the Beehive in Cancer, Berenice's Hair, the Hyades, and the group in the sword-handle of Perseus. The stars of which they are composed can generally be easily distinguished by

Fig. 82.

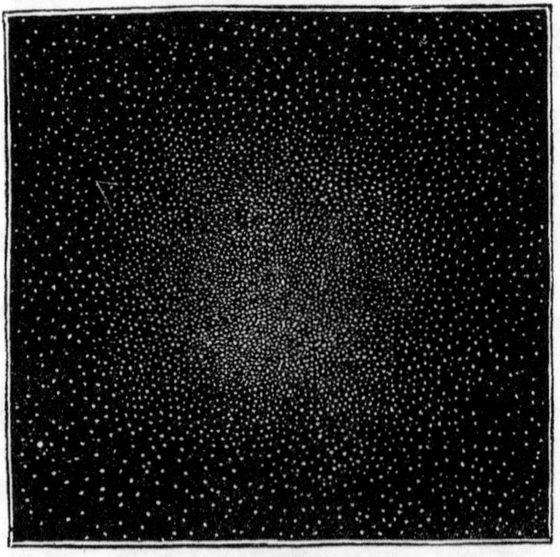

STAR-CLUSTER IN TOUCAN.

the naked eye, although by the use of a small opera or spy glass the number is largely increased. In the southern sky there are clusters still more remarkable. In the Cross is a group of 110 stars of various colors, red, blue, and green, so that looking on it, says Herschel, is "like gazing into a casket of precious gems." A cluster in Toucan is compact at the centre, where it is of an orange-red color; the exterior is composed of pure white stars, making a border of exquisite contrast. It is generally conceded that there is some close physical relation existing between the stars composing such an "archipelago of worlds," but its nature is a mystery. They seem generally crowded together toward the centre, blending into a continuous blaze of light. Yet, although they appear so densely compacted, it is probable that if we could change our stand-point, penetrating one of these groups of suns we should find it opening up and spreading out before us on our approach, until, in the midst, the suns would shine down upon us from the heavens as the stars do in our own sky.

NEBULÆ.—These are faint misty objects like specks of luminous clouds. They are generally either round or oval, and brightest at the centre. They differ from "clusters" in not being resolvable into stars when viewed through the largest telescopes. With the constant improvement made in these instruments, however, many nebulæ have been resolved, and thus the number of clusters increased, while new nebulæ are being discovered to take their places. Until of late, it was thought that all nebulæ were simply groups of stars, which would be ultimately discerned in the more powerful telescopes yet to be made. Spectrum analysis shows, however, that many of these luminous clouds are gaseous, and not solid. They cannot, therefore, be suns. Since they maintain the same position with respect to the stars, their distance must be inconceivably great, and in order to be visible to us, their magnitude must be proportionately vast. They are most abundant at the two poles of the Milky Way, but are more uniformly distributed over the heavens lying near the south pole. Those portions of the sky which are poorest in stars, are richest in nebulæ. Herschel was accustomed to say to his secretary, whenever for a brief time he saw no star passing the field of his telescope, as in the diurnal revolution the heavens swept by it, "Prepare to write; nebulæ are about to arrive."

Nebulæ are divided, according to their form, into six classes—*elliptic*, *annular*, *spiral*, *planetary*, *irregular nebulæ*, and *nebulous stars*.

The elliptic, or merely oval nebulæ, are the most abundant. Under this head is commonly classed the "great nebula in Andromeda," which was discovered over a thousand years ago. It is visible to the naked eye. Prof. Bond, of the Cambridge Observatory, has partly resolved it into stars. He has distinctly counted 1500, although its nebulous appearance is still retained. Under the telescope it is one of the

Fig. 83.

NEBULA IN ANDROMEDA.

most glorious objects in the heavens. "If we suppose this nebula to be one continuous bed of stars of different sizes for its entire extent, it must comprise the enormous number of 30,000,000." The distance of such nebulæ from the earth entirely passes our comprehension. Some astronomers have estimated that a ray of light would require 800,000 years to span the gulf that intervenes. Imagination wearies itself in the attempt to understand these figures. They only teach us something of the limitless expanses of that space in which God is working the mysterious problem of creation.

The *annular nebulæ* have the form of a ring. There are but four of these "ring universes." In

Fig. 84.

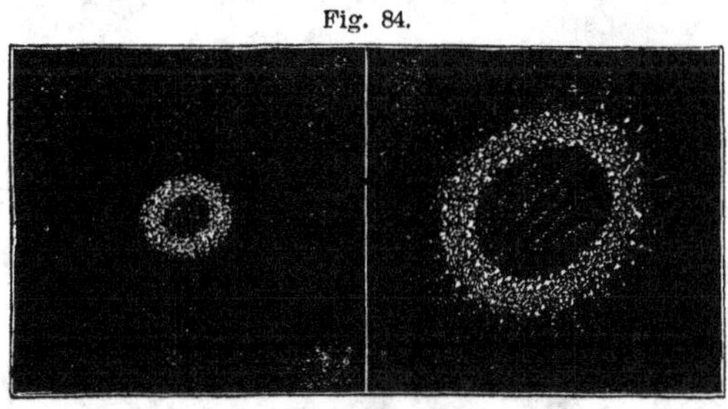

NEBULA IN LYRA.

the cut is a representation of one in Lyra—first as seen by Herschel, and having in the centre a nebulous film like a "bit of gauze stretched over a hoop;" second, as shown in Lord Rosse's great telescope, which resolves the filmy parts of the nebula into excessively minute stars, and reveals a fringe of stars

Fig. 85.

SPIRAL CLUSTER IN CANES VENATICI.

along the edge. Though apparently so small, its dimensions must be enormous. If no further from the earth than 61 Cygni, the diameter would be 2,000,000,000 miles. It is probably immensely further distant.

The *spiral* or "whirlpool nebulæ" are exceedingly curious in their appearance. The most remarkable one is that in Canes Venatici. It consists of brilliant spirals sweeping outward from a central nucleus, and all overspread with a multitude of stars. One is lost in attempting to imagine the distance of such a mass, and the forces which produce such a "tremendous hurricane of matter—perhaps of suns."

Planetary nebulæ, by their circular form and pale uniform light, resemble the disks of the most distant planets of our system. Their edges are generally well defined, though sometimes slightly furred. Three-fourths of them are in the southern hemisphere. Several have a blue tinge. There is one in Ursa Major, which if located at the distance named before—that of 61 Cygni—would fill a space equal to three times the entire orbit of Neptune. About twenty-five of these "island universes" have been found scattered through the ocean of space. Columbus discovered a new continent, and so immortalized his name; what shall we

Fig. 86.

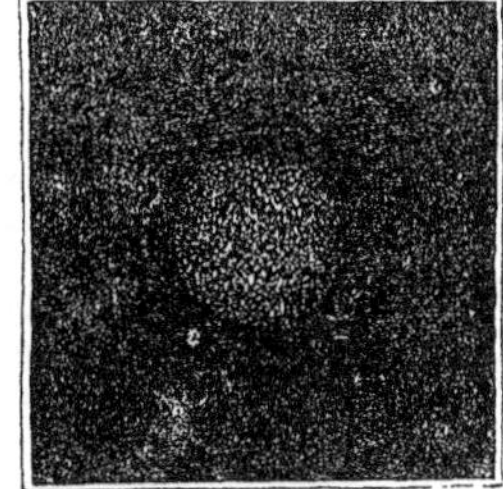

PLANETARY NEBULA.

say of the astronomer who discovers a universe of worlds?

Irregular nebulæ are those which have no definite form. Many of them present all the irregularities of clouds torn and rent by the tempest. Some of the likenesses which may be traced by the fancy are strangely fantastic: for example, the "dumb-bell nebula" in the constellation Vulpecula, and the "crab nebula" near the southern horn of Taurus. There is also one known as "the great nebula in the sword-handle of Orion," in which may be seen a faint resemblance to the wings of a bird.

Fig. 87.

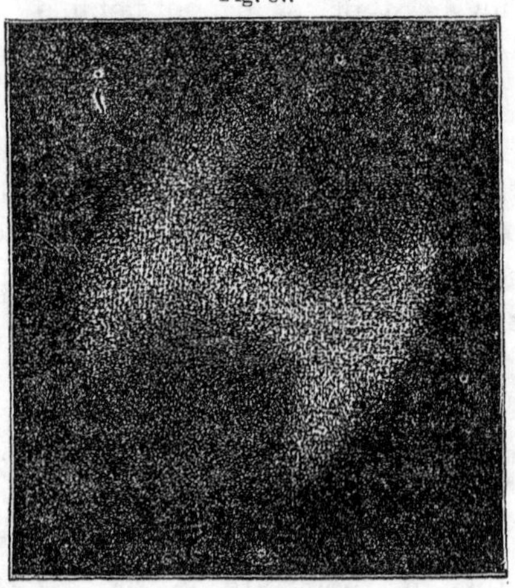

DUMB-BELL NEBULA.

Nebulous stars are so called because they are enveloped by a faint nebula, usually of a circular form. The star is generally seen at the centre, although some which are elliptical surround two stars, one in each focus. It is thought that these may be suns possessing immense atmospheres, which are rendered visible somewhat as that of our sun is in the zodiacal light; and that in like manner our sun

Fig. 88.

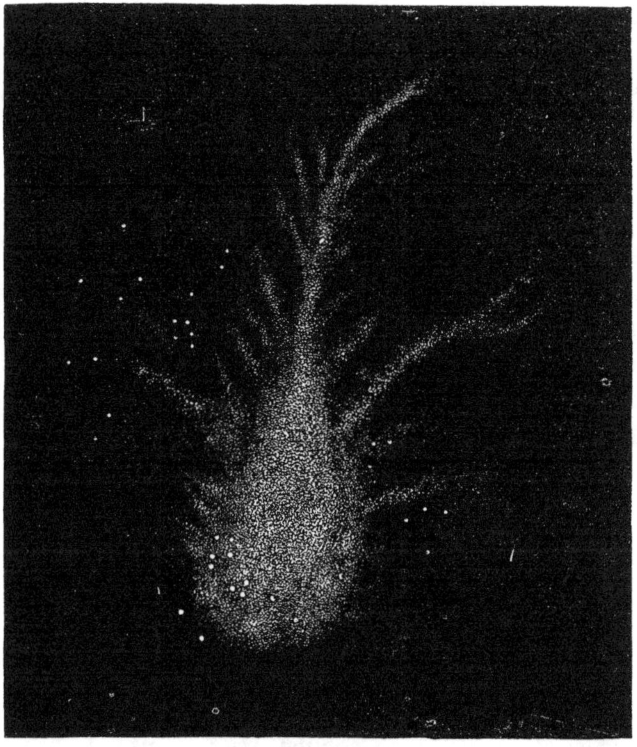

CRAB NEBULA.

itself to those in space presents the appearance of a nebulous star. The luminous atmosphere of the star in Cygnus, if located at the distance of α Centauri, is of an extent equal to "fifteen times the distance of Neptune from the sun."

Variable nebulæ.—Certain changes take place among the nebulæ which can be accounted for only under the supposition that they, like some of the stars, are *variable.* Mr. Hind tells us of one in Taurus which was distinctly visible with a good telescope in 1852, but in 1862 it had vanished entirely out of the reach of a much more powerful instrument. It seems to have disappeared altogether. The great nebula in Argo, when observed by Herschel in 1838, had in the centre a vacant space containing a star of the first magnitude completely enshrouded by nebulous matter. In 1863, the nebulous matter had disappeared, and the star was only of the sixth magnitude. These facts as yet defy explanation. They only illustrate the vast and wonderful changes constantly taking place in the heavens.

Double nebulæ.—There seems to be a physical connection existing between some of the nebulæ, similar to that already noticed in respect to certain stars. In the case of the latter, this inter-relation has been proved, since their movements even at their distances can yet be traced in the lapse of years. "But owing to the almost infinite depths in the abyss of the heavens at which these nebulæ exist, thousands of years, perhaps thousands of centuries, would be necessary to reveal any movement." (Guillemin.)

MAGELLANIC CLOUDS.—Not far from the southern pole of the heavens there are two cloud-like masses, distinctly visible to the naked eye, known to navigators as "Cape Clouds." Sir John Herschel describes them as consisting of swarms of stars, clusters, and nebulæ, seemingly grouped together in the wildest confusion. In the larger, he found 582 single stars, 46 clusters, and 291 nebulæ.

THE MILKY WAY.—Via Lactea or Galaxy, as it is variously termed—is that luminous, cloud-like band that stretches across the heavens in a great circle. It is inclined to the celestial equator about 63°, and intersects it in the constellations Cetus and Virgo. This stream of suns is divided into two branches from α Centauri to Cygnus. To the naked eye it presents merely a diffused light; but with a powerful telescope it is found to consist of myriads of stars densely crowded together. These stars are not uniformly distributed through its entire extent. In some regions, within the space of a single square degree we can discern as many as can be seen with the naked eye in the entire heavens. In other parts there are broad open spaces. A remarkable instance of this occurs near the Southern Cross. There is a dark pear-shaped vacancy with a single bright star at the centre, glittering on the blue background of the sky. In viewing it, one is said to be impressed with the idea that he is looking through an opening into the starless depths beyond the Milky Way.

The number of stars in the galaxy which may be seen by Herschel's great reflector is estimated at twenty-one and a half millions. With the more powerful instruments now being made it is probable the number will be largely increased. The northern galactic pole is situated near Coma Berenices, and the southern in Cetus. Advancing from either pole toward the Milky Way, the number of stars increases, at first slowly and then more rapidly, until the proportion at the galaxy itself is thirty-fold.

Herschel's theory.—Sir W. Herschel has conjectured that the stars are not indifferently scattered through space, but are collected in a stratum something like that shown in the cut, and that our sun

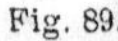
Fig. 89.

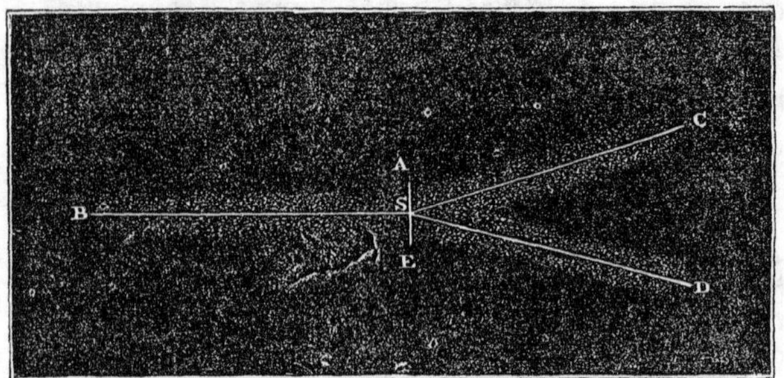

HERSCHEL'S THEORY OF THE MILKY WAY.

occupies a place at S, near where the stream branches. A and E are the galactic poles. It is evident that, to an eye viewing the stratum of stars in the direction SB, SC, or SD, they would seem much denser than in the direction SA or SE. Thus are we to think of our own sun as a star of the second or third magnitude, and our little solar system as plunged far into the midst of this vortex of worlds, a mere atom along that

> "Broad and ample road
> Whose dust is gold and pavement stars."

NEBULAR HYPOTHESIS.—This is a theory which was advanced by Laplace, to show how the solar system was formed. In the "beginning," all the matter which now composes the sun and the various planets, with their moons, was in a gaseous and highly heated state. It filled all the space now occupied by the system, and extended far beyond the orbit of Neptune. In other words, the solar system was simply an immense nebula. The heat, which is the repellant force, overcame the attraction of gravitation. Gradually the mass cooled by radiation. As centuries passed, the repellent force becoming weaker, the attractive force drew the matter and condensed it toward one or more centres. The nebula then presented the appearance of a nebulous star—a nucleus enveloped to a great distance by a gaseous atmosphere. According to a well-known law in philosophy, seen in every-day life, in a whirlpool, a whirlwind, or even in water poured into a funnel, wherever matter seeks a centre, a rotary motion is established. As this rotary motion increased, the centrifugal force finally overcame at the exterior the attraction of gravitation, and so threw off a ring of condensed vapor. Centuries elapsed, and again, under the same conditions, a second ring was detached. Thus, one by one, concentric rings were separated from the parent nebula, all revolving in the same plane and in the same direction. These different rings, becoming gradually consolidated, formed the planets,—generally however, in this process, while still in the vaporous state and slowly condensing, themselves throwing off rings which were in turn consolidated into satellites. In the case of Saturn, several of these secondary rings did not break up, and so condense into globes, but still remain as rings which revolve about the planet.* Mitchell naïvely remarks, "Saturn's rings were left unfinished to show us how the world was made." The ring which formed the minor planets broke up into small fragments, none large enough to attract the rest and thus form a single globe. The central mass of vapor finally condensed itself into the sun, which remains the largest member of the system. According to this theory, the sun may yet give off a few more planets, whose orbits will not exceed its present diameter. After a time its heat will have all been radiated into space, its fire will become extinct, and life on the planets will cease. We know not when this remote event may occur. We cannot fathom the purpose of God in creating and maintaining this system of worlds, nor foretell how soon it may complete its mission. We are assured, however,

"That nothing walks with aimless feet,
That not one life shall be destroyed,
Or cast as rubbish to the void,
When God hath made the pile complete."

In Memoriam.

---

* It is possible that these rings may yet break up and form new satellites for that planet. Indeed, some hold that one at least of the rings has thus been resolved into small meteorites. These may be attracted, and so picked up, one by one, by the larger in succession, until they form another moon, which will continue to revolve about the planet as the ring does now.

Fig. 90.

CELESTIAL CHEMISTRY.

Spectrum Analysis.—The rainbow—that child of the sun and shower—is familiar to all. The brilliant band of colors, seen when the sunbeam is passed through a prism, is scarcely less beautiful. The ray of light containing the primary colors is spread out fan-like, and each tint reveals itself. This variously colored band is called in philosophy a spectrum (plural, *spectra*). There are three different kinds of spectra—

1st. When the light of a solid or liquid body, as

iron white-hot, is passed through a prism, the *spectrum is continuous*, and consists of a series of distinct colors, varying from red on one side to violet on the other.

2d. If the light of a burning gas containing any volatilized substance be passed through a prism, the *spectrum is not continuous*, but is ornamented by bright-colored lines—sodium giving two yellow lines, strontia a red one, silver two beautiful green ones. Each element produces a definite series which can be readily recognized as its test.

3d. If a light of the first kind be passed through one of the second, the spectrum will be found to be crossed by *dark lines*. Thus, if the white light of a burning match be passed through a flame containing sodium, instead of the vivid yellow lines so characteristic of that metal, two black lines will exactly occupy their place. *A gaseous flame absorbs the rays of the same color that it emits.*

THE SPECTROSCOPE.—This instrument consists of two small telescopes, with a prism mounted between their object-glasses, in the manner shown in the cut. The rays of light enter through a narrow slit at A and are rendered parallel by the object-glass. They then pass through the prisms at C, are separated into the different colors, and entering the second telescope at D, fall upon the eye at B. A third telescope is sometimes attached, which contains a minutely accurate scale for measuring the distances of the lines. In addition, a mirror may throw in a ray of sunlight or starlight at one side of the slit, and so we can compare the spectrum of the sunbeam with that of any flame we desire.

Fig. 91

A SPECTROSCOPE.

Revelations of the spectroscope concerning the sun.—The spectrum of the sunbeam is not continuous, but is crossed by a large number of dark lines, called, from their discoverer, Fraunhofer's lines. It is therefore concluded that the sun's light is of the third class just named, and that it is produced by the vivid light of a highly heated body shining through a flame full of volatilized substances. But not only does spectrum analysis thus shed light on the physical constitution of the sun, but these lines are so distinctive, so marked and varied, that the very elements of which the sun is composed may be discovered. Thus, for example, iron gives a spectrum of some 70 lines, differing in intensity and relative length. These are bright when iron vapor is burning, and dark when white light is passed through such burning vapor. In the solar spectrum we have the perfect coincidence of 70 dark lines, line for line and strength for strength. The conclusion is irresistible that iron is contained in the sun's atmosphere. The following include all the elements that are now known to exist in it:

Sodium,	Iron,	Strontium
Calcium,	Chromium,	Cadmium,
Barium,	Nickel,	Cobalt,
Magnesium,	Zinc,	Hydrogen.

STARS ARE SUNS.—The same method of analysis has been applied to the stars. Their spectra also are marked by dark lines. Their constitution is therefore like our sun; they contain also the same familiar elements. Aldebaran seems the most like our earth. It has at least nine elements known to chemists:

Sodium,	Iron,	Magnesium,
Hydrogen,	Bismuth,	Antimony,
Tellurium,	Mercury,	Calcium.

Betelgeuse contains many elements known to us, but no hydrogen.—What a world that must be without water! Arcturus, Rutherford says, closely resembles our sun.

We thus trace in the faintest star that trembles in the measureless depths of space the same elements that compose the food we eat and the water we drink. We know that we are akin to nature everywhere—that we are a part of a system vast as the universe.

SPECTRA OF NEBULÆ.—Instead of being marked with dark lines, as are the spectra of the stars, many of these exhibit bright lines. Their spectra are of the 2d kind. This proves the nebula to consist, not, like the stars, of an intensely heated solid body shining through a luminous atmosphere, but of a glowing mass of gas. Out of 60 nebulæ examined by Mr. Huggins, 20 exhibited the bright lines belonging to the gases, and all contained nitrogen.

It is possible in this manner even to decide the relative brightness of the different nebulæ. The dumb-bell nebula was found to emit a light only about one twenty-thousandth part that of a common wax-candle. If this matter be a "sun-germ," how immensely must it become condensed before its rushlight glimmering can rival the dazzling brilliancy of even our own sun!

THE SOLAR FLAMES, which before were seen only at an eclipse, can now be examined at any time. The sun is a sea of fire. Flames travel over its surface faster than the earth in its orbit: one shot out 80,000 miles and disappeared in ten minutes. Such tremendous convulsions surpass all terrestrial phenomena.

TIME.

SIDEREAL TIME.—A sidereal day is the exact interval of time in which the earth revolves on its axis. It is found by marking two successive passages of a star across the meridian of any place. This is so absolutely uniform, that the length of the sidereal day has not varied $\frac{1}{100}$ of a second in 2,000 years. The sidereal day is divided into twenty-four equal portions, which are called sidereal hours, and each of these into sixty portions, termed sidereal minutes, etc.

Astronomical clocks are regulated to keep sidereal time. The day commences when the vernal equinox is on the meridian. Therefore, the time by the sidereal clock does not in any way point out the hour of the ordinary day. It only indicates how long it is since the vernal equinox crossed the meridian, and thus always shows the right ascension of any star which may happen to be on the meridian at that moment. The hours of the clock are easily reduced to degrees (see. p. 38). The astronomer always reckons the hours of the day consecutively up to twenty-four.

Solar Time.—A solar day is the interval between two successive passages of the sun across the meridian of any place. If the earth were stationary in its orbit, the solar day would be of the same length as the sidereal; but while the earth is turning around on its axis, it is going forward at the rate of 360° in a year, or about 1° per day. When the earth has made a complete revolution, it must therefore perform a part of another revolution through this additional degree, in order to bring the same meridian vertically under the sun. One degree of diurnal revolution is about equal to four minutes of time.

Hence the solar day is about four minutes longer than the sidereal day. For the convenience of society, it is customary to call the solar day 24 hours long, and make the sidereal day only 23 hr. 56 min. 4 sec. in length, expressed in mean solar time. A sidereal day being shorter than a solar one, the sidereal hours, minutes, etc., are shorter than the solar; 24 hours of mean solar time being equal to 24 hr. 3 min. 56 sec. of sidereal time.

From what has been said, it follows that the earth makes 366 revolutions around its axis in 365 solar days.

MEAN SOLAR TIME.—The solar days are of unequal length. To obviate this difficulty, astronomers suppose a *mean sun* moving through the equator of the heavens (which is a circle and not an ellipse) with a perfectly uniform motion. When this mean sun passes the meridian of any place, it is *mean noon*; and when the true sun is in the same position, it is *apparent noon.* This day is the average length of all the solar days in the year. The clocks in common use are regulated to keep mean time. When, therefore, it is twelve by the clock, the sun may be either a little past or a little behind the meridian. The difference between the sun-time (apparent solar-time) and the clock-time (mean time), is called the "*equation of time.*" This is the greatest about the first of November, when the sun is sixteen and a quarter minutes in advance of the clock. The sun is the slowest about February 10th, when it is about fourteen and a half minutes behind mean time. Mean and apparent time coincide four times in the year—namely, April 15th, June 15th, September 1st, and December 24th. On those days the noon-mark on the sun-dial coincides with twelve o'clock. In France, until 1816, apparent time was used; and the confusion was so great, that Arago relates how the town clocks would differ thirty minutes in striking the same hour. As the time varied every day, no watchmaker could regulate a watch or clock to keep it.

THE SUN-DIAL.—The *apparent time* of the dial may be readily changed to mean time, by adding or subtracting the number of minutes given in the almanac for each day in the year, under the heading "sun slow" or "sun fast." As a noon-mark is thus a very convenient method of regulating a timepiece, especially in the country, the following manner of obtaining one without a transit instrument may be useful.

Select a level hard surface which is exposed to the sun from about 9 A.M. to 3 P.M. Upon this carefully describe, with compasses, a circle of eight or ten inches in diameter. Take a piece of heavy wire, six or eight inches in length, one end of which is sharpened. Drive this *perpendicularly* into the centre of the circle, leaving it just high enough to allow the extreme end of its shadow to fall upon the circle about 9½ or 10 A.M. Mark this point, and also the place where the shadow touches the circle in the afternoon. Take a point half-way between the two, and drawing a line from that to the centre of the circle, it will be the *meridian line* or noon-mark.

WHY THE SOLAR DAYS ARE OF UNEQUAL LENGTH.—There are two reasons for this—the unequal orbital motion of the earth and the obliquity of the ecliptic. First: the orbit of the earth is an ellipse; and thus the apparent yearly motion of the sun along the ecliptic is variable. In perihelion, in January, the sun appears to move eastward daily 1° 1′ 9.9″; while at aphelion, in July, only 57′ 11.5″. As the earth in its diurnal motion revolves *uniformly* from west to east, and the sun passes eastward *irregularly*, this must produce a corresponding variation in the length of the solar day. The sun, therefore, comes to the meridian sometimes earlier and sometimes later than the mean noon, and they agree only at perihelion and aphelion.

Second: as we have just seen, the mean sun is supposed to move in a circle and not an ellipse. This would make the motion along the ecliptic uniform, but the obliquity of the ecliptic would still cause an irregularity in the length of the day. The mean sun is therefore supposed to pass along the equinoctial,

Fig. 92.

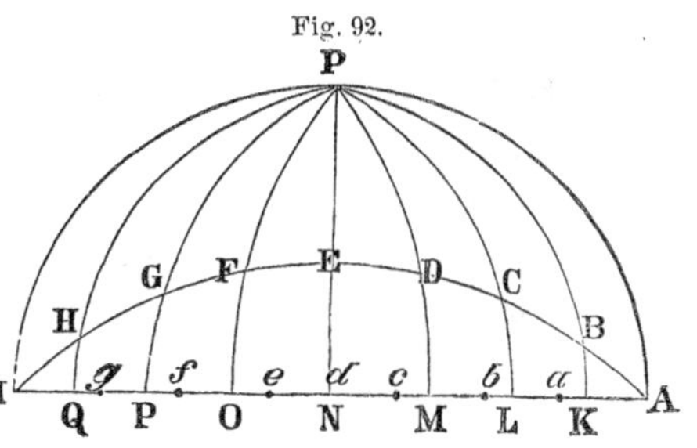

which is perpendicular to the earth's axis; while the ecliptic is inclined to it 23° 28′. Let A represent the vernal equinox, I the autumnal, AEI the ecliptic, AI the equinoctial, PK, PL, PM, etc., meridians. Let the distances AB, BC, CD, etc., be equal arcs of the ecliptic, which are passed over by the sun in equal times. Next, mark off on the equinoctial distances A*a*, *ab*, *bc*, etc., equal to AB, BC, etc. These are equal arcs of right ascension, or hour-circles, through which the earth, revolving from west to east, passes in equal times. Now, meridians drawn through these divisions, would not agree with those drawn through equal divisions on the ecliptic. Hence, a sun moving along the ecliptic, which is inclined, would not make equal days, even though the ecliptic were a perfect circle. Let us see how the mean and apparent solar days would compare. Let the real sun pass in its eastward course from A to B in a certain time, the mean sun moving the same distance would reach the point *a*, since the latter travels on the base and the former the hypothenuse of a triangle. The earth, revolving from west to east, would cause the real sun to cross any meridian earlier than the mean sun; hence, apparent time would be faster than clock-time. By holding the figure up above us toward the heavens, we can see how a westerly sun would cross the meridian earlier than an easterly one. Following the same reasoning, we can see that at the solstice, solar and mean time would agree; while beyond that point the mean time would be faster.

THE CIVIL DAY.—This is the mean solar day of which we have spoken. It extends from midnight to midnight. The present method of dividing the day into two portions of twelve hours each, was adopted by Hipparchus, 150 years B. C., and is now in general use over the civilized world. Until recently, however, very many nations terminated one day and commenced the next at sunset. Under this plan, 10 o'clock on one day would not mean the same as 10 o'clock on another day. The Puritans commenced the day at 6 P. M. The Babylonians, Persians, and modern Greeks begin the day at sunrise. The names of the days now in use are derived as follows:

1. Dies Solis.......(Latin)....Sun's day.
2. Dies Lunæ.....(")....Moon's day.
3. Tius daeg.......(Saxon)....Tius's day.
4. Wodnes daeg...(")....Woden's day.
5. Thurnes daeg..(")....Thor's day.
6. Friges daeg ....(")....Friga's day.
7. Dies Saturni...(Latin)....Saturn's day.

THE YEAR.—The *sidereal year* is the interval of a complete revolution of the earth about the sun, measured by a fixed star. It comprises 365 d., 6 hrs., 9 min., 9.6 sec. of mean solar time. The *mean solar year* (tropical year) is the interval between two successive passages of the sun through the vernal equinox. It comprises 365 d., 5 hrs., 48 min., 49.7 sec. If the equinoxes were stationary, there would be no difference between the sidereal and tropical year. As the equinoxes retrograde along the ecliptic 50″ of space annually, the former is 20 min., 20 sec. longer.

The *anomalistic year* is the interval between two successive passages of the earth through its perihelion. It is 4 min., 40 sec. longer than the sidereal year.

THE ANCIENT YEAR.—The ancients ascertained the length of the year by means of the *gnomon*. This was a perpendicular rod standing on a smooth plane on which was a meridian line. When the shadow cast on this line was the shortest, it indicated the summer solstice; and when it was the longest, the winter solstice. The number of days required for the sun to pass from one solstice back to it again determined the length of the year. This they found to be 365 days. As that is nearly six hours less than the true solar year, dates were soon thrown into confusion. If, at a certain date the summer solstice occurred on the 20th June, in four years it would fall on the 21st; and thus it would gain one day every four years, until in time the summer solstice would happen in the winter months.

JULIAN CALENDAR.—Julius Cæsar first attempted to make the calendar year coincide with the motions of the sun. By the aid of Sosigenes, an Egyptian astronomer, he devised a plan of introducing every fourth year a leap-year, which should contain an extra day. This was termed a *bissextile year*, since the sixth (sextilis) day before the kalends (first day) of March was then counted twice.

GREGORIAN CALENDAR.—Though the Julian calendar was nearly perfect, it was yet somewhat defective. It considered the year to consist of 365¼ days, which is 11 min. in excess. This accumulated year by year, until in 1582 the difference amounted to ten days. In that year, the vernal equinox occurred on the 11th of March, instead of the 21st. Pope Gregory undertook to reform the anomaly, by dropping ten days from the calendar and ordering that thereafter only centennial years which are divisible by 400 should be leap-years. The Gregorian calendar was generally adopted in all Catholic countries. Protestant England did not accept the change until 1752. The difference had then amounted to 11 days. These were suppressed and the 3d of September was styled the 14th.

Dates reckoned according to the Julian calendar are termed Old Style (O. S.), and those according to the Gregorian calendar New Style (N. S.) This sweeping change was received in England with great dissatisfaction. Prof. De Morgan narrates the following. "A worthy couple in a country town, scandalized by the change of the calendar, continued for many years to attempt the observance of Good Friday on the old day. To this end they walked seriously and in full dress to the church door, on which the gentleman rapped with his stick. On finding no admittance, they walked as seriously back again and read the service at home. There was a wide-spread superstition that, when Christmas day began, the cattle fell on their knees in their stables. It was asserted that, refusing to change, they continued their prostrations according to the Old Style. In England, the members of the government were mobbed in the streets by the crowd, which demanded the eleven days of which they had been illegally deprived."

COMMENCEMENT OF THE YEAR.—The Jews began their civil year with the autumnal equinox, but their ecclesiastical with the vernal. When Cæsar revised the calendar, among the Romans the year commenced with the winter solstice (Dec. 22), and it is probable he did not intend to change it materially. He, however, ordered it to date from January 1st, in order that the first year of his new calendar should begin with the day of the new moon immediately succeeding the winter solstice.

THE EARTH OUR TIMEPIECE.—The measure of time is, as we have just seen, the length of the mean day. That is estimated from the length of the sidereal day. Hence the standard for time is the revolution of the earth on its axis. All weights and measures are based on time. An ounce is the weight of a given bulk of distilled water. This is measured by cubic inches. The inch is a definite part of the length of a pendulum which vibrates seconds in the latitude of London. Arago remarks, a man would be considered a maniac who should speak of the influence of Jupiter's moons on the cotton trade. Yet there is a connection between these incongruous ideas. The navigator, travelling the waste of waters where there are no paths and no guide-boards, may reckon his longitude by the eclipses of Jupiter's moons, and so decide the fate of his voyage. We can easily see how the revolution of the earth on its axis influences the cost of a cup of tea.

CELESTIAL MEASUREMENTS.

Many persons read the enormous figures which indicate the distances and dimensions of the heavenly bodies with an indefinite idea, which conveys no such feeling of certainty as is experienced when they read of the distance between two cities, or the number of square miles in a certain State. Many, too, imagine that celestial measurements are so mysterious in themselves that no common mind can hope to grasp the methods. Let us attempt the solution of a few of these problems.

1st. TO FIND THE DISTANCES OF THE PLANETS FROM THE SUN.—In the figure, E represents the earth, ES the earth's distance from the sun, V the planet Venus, and VES the angle of elongation (a right-angled triangle). It is clear, that as Venus swings apparently east and west of the sun, this angle may be easily measured; also, that it will be the greatest when Venus is in aphelion and the earth

Fig. 93.

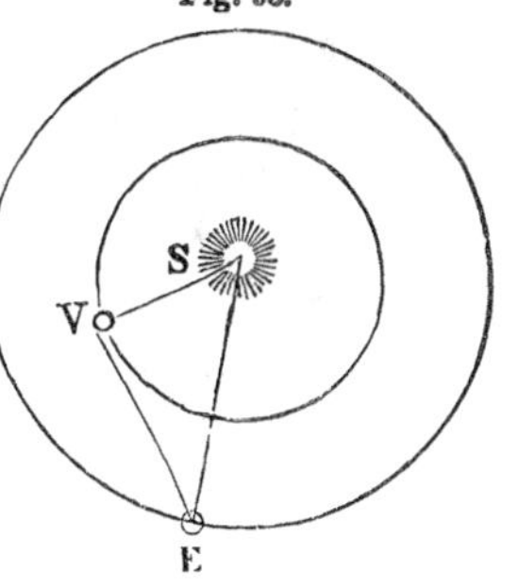

COMPARATIVE DISTANCE OF VENUS AND THE EARTH.

in perihelion at the same time, for then VS will be the longest and VE the shortest. Now in every right-angled triangle the proportion between the hypothenuse, ES, and the side opposite, VS, changes as the angle at E varies, but with the same angle remains the same whatever may be the length of the lines themselves. This proportion between the hypothenuse and the side opposite any angle is termed the *sine of that angle.* Tables are published which contain the sines for all angles. In this way, the mean distance of Venus is found to be $\frac{72}{100}$* that of the earth, Mars $\frac{4}{3}$ times, Jupiter $5\frac{1}{5}$ times, etc.

The same result would be obtained by the use of Kepler's third law; and on page 29, we saw how the distances of the planets themselves could be determined by the periodic times, if the distance of the earth from the sun is first known. So that when we have accurately determined the sun's distance from us, we can then decide by either of the methods named the distance of all the planets. Indeed that is, as already remarked, the "foot-rule" for measuring all celestial distances.

2d. TO MEASURE THE MOON'S DISTANCE FROM THE EARTH.—(1.) *The ancient method.*—As the moon's distance is so much less than that of the other heavenly bodies, it is measured by the earth's semi-diameter.

---

* If the pupil has studied Trigonometry, he may apply here the simple proportion—

ES : VS :: Radius : Sine of 47° 15″ = greatest elongation of Venus.

The method, an extremely rough one, which was in use among the ancients, was something like the following. In an eclipse of the moon, that body passes through the earth's shadow in about four hours. If, then, the moon travels along its orbit in four hours a distance equal to the diameter of the earth, in twenty-four hours it would pass over six times, and in a lunar month (about thirty days) one hundred and eighty times, that distance. The circumference of the lunar orbit must be then one hundred and eighty times the diameter of the earth. The ancients supposed the heavenly orbits to be circles, and as the diameter of a circle is about $\frac{1}{3}$ of the circumference, they deduced directly the diameter of the moon's orbit as 120 times, and the distance of the the moon from the earth as 60 times the semi-diameter of the earth.

(2.) *Modern method by the lunar parallax.*—Under the head of parallax we saw how, in common life, we obtain a correct idea of the distance of an object by means of our two eyes. We proved that one eye alone gives no notion of distance. Just, then, as we use two eyes to find how far from us an object is, so the astronomer uses two astronomical eyes or observatories, located as far apart as possible, to find the parallax of a heavenly body. In the figure, M represents the moon, G an observatory at Greenwich, and C another at the Cape of Good Hope. At the former, the distance from the north pole to the centre of the moon, measured on a meridian of the celestial sphere, is found to be 108°. At the latter station, the distance from the south pole to the moon's centre is measured in the same way, and found to be 73½°. The sum of these angles is 181½°. Now, the entire distance from the north pole around to the south pole, measured on a meridian, can be only half a great circle, or 180°. This difference of

Fig. 94.

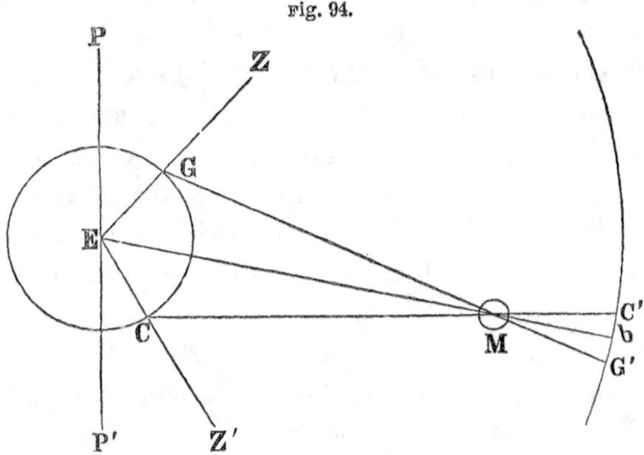

MEASURING MOON'S DISTANCE FROM THE EARTH.

1½° must be the difference in the position of the moon, as seen from the two observatories. For the observer at the former station will see the moon projected on the celestial sphere at G′, and in measuring its distance from the north pole will measure an arc *b*G′ further than if he were located at E, the centre of the earth. The observer at the latter station will see the moon projected on the celestial sphere at C′, and in measuring its distance from the south pole will measure an arc bC' more than if he were located at E, the centre of the earth. The sum of bG' and $bC' = G'C'$ is the difference in the position of the moon as seen from the two stations. In other words, it is the moon's parallax. The arc $G'C'$ measures the angle $C'MG'$; that angle is equal to the opposite angle $GMC = 1\frac{1}{2}°$. Now, in the four-sided figure GECM, the sides GE and CE are each equal radii of the earth = 3956 miles; while the distance from G to C is the difference in the latitude of the two places. The angles ZGM and $Z'CM$, being the zenith distances of the moon, are known, and so the angles MGE and MCE are easily found. EM, the moon's distance from the centre of the earth, is thus readily computed by a simple trigonometrical formula.

(3.) *The horizontal parallax of the moon* is most commonly found by estimating its distance, not from the north and south poles, as just explained under the general meaning of the term parallax, but from a fixed star. The moon's horizontal parallax is now estimated at 57′, which makes its distance about sixty times the earth's semi-diameter.*

TO FIND THE SUN'S DISTANCE FROM THE EARTH.— This might be estimated by obtaining the solar

---

* In figure 95, let S represent the moon, sun, or any other heavenly body, AB the semi-diameter of the earth, and ASB the "horizontal parallax" of the body. Then, by the following trigonometrical formula, the distance from the earth may be easily calculated—

AS : AB :: Radius : Sin of ASB.

parallax in the same manner as the lunar parallax. It would be only necessary to take the sun's distance from the north and south poles respectively at Greenwich and the Cape of Good Hope, and then subtracting 180° from the sum of the two angular distances, the remainder would be the solar parallax. The difficulty in this method lies in the fact that when the sun shines the air is full of tremulous motion. This increases refraction—that plague of all astronomical calculations—to such an extent that it becomes impossible to calculate so small an angle with any accuracy. Neither can the parallax be estimated, as in the case of the moon, by measuring

Fig. 95.

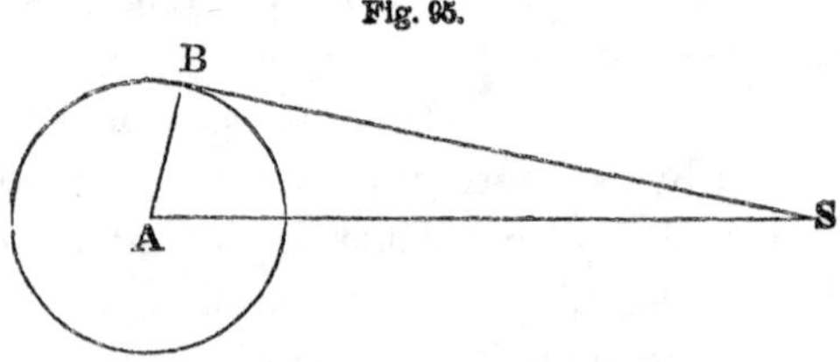

the distance from a fixed star, since when the sun shines the stars near by are invisible even in a telescope. Astronomers have therefore been compelled to resort to other methods.

(1.) *Calculation of solar parallax by observation of the planet Mars.*—We have already seen that the distance of Mars from the sun is $\frac{4}{3}$ that of the earth from the sun. If, therefore, we can find Mars' distance from the earth, we can multiply it by three, and so obtain the distance of the sun from the earth. In 1862, when Mars was in opposition, it came very near us, for it was in perihelion while the earth was in aphelion, so that its distance (as since ascertained) was only 126,300,000 – 93,000,000 = 33,300,000 miles. Observers at Greenwich and the Cape, and at various American and European observatories, calculated the distance of the planet from the north and south poles, as well as from several fixed stars, in precisely the manner just explained for obtaining the lunar parallax. The result of these observations fixed the solar parallax at 8.94″.*

(2.) *Calculation of solar parallax by observation of the transit of Venus.*—In the figure, let A and B represent the positions of two observers stationed at

Fig. 96.

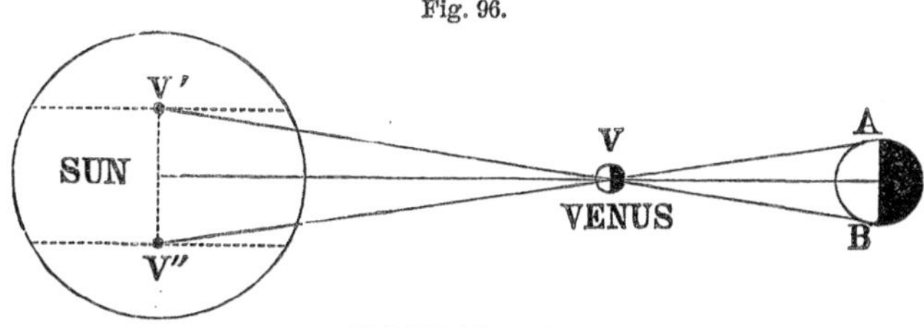

TRANSIT OF VENUS.

opposite sides of the earth. At the time of the transit, the one at A will see the planet Venus as a round black spot at V″ on the sun's disk, while the one at B will see it at V′. The distance V′V″ is the

* By the formula on page 302, we have—

AS : AB :: Radius : Sin 8.94″.

difference in the position of Venus as seen from the two stations on the earth. The distance AB is the diameter of the earth. The distance V′V″ is as much greater than AB as VV″ is greater than VA. The distance of Venus from the sun is known, by Prob. I., to be .72 that of the earth. The distance of Venus from the earth must be, then, $1.00 - .72 = .28$. Hence, VV″, the distance from the sun to Venus, $= .72 \div .28 = 2.5$ times the length of AV, the distance of Venus from the earth. Therefore, V′V″ is equal to $2\frac{1}{2}$ times AB, the earth's diameter, or 5 times the solar parallax. Knowing the hourly motion of Venus, it is only necessary for each observer to find when the planet's disk enters upon and leaves the sun's disk to determine the length of the path (*chord*) it traces. A comparison of the length and direction of these chords will give the length of V′ V″ in seconds of space.

The advantage of this method is, that as the distance V′ V″ is two and a half times that of AB, an error in measuring that chord affects the solar parallax less than one-fifth.

Time of a transit of Venus.—This is an event of rare occurrence. It happens only at intervals of 8, $105\frac{1}{2}$; 8, $121\frac{1}{2}$, years, &c. Were the planet's orbit in the same plane as the ecliptic, a transit would take place during each synodic revolution; but as it is inclined about $3\frac{1}{2}°$, the transit can occur only when the earth is at or near one of the nodes at the same time with the planet, when in inferior conjunction. As the nodes of Venus fall in that part of the earth's orbit which we pass in the beginning of June and December transits always occur in those months.

The transit of June 3d, 1769, excited great interest. King George III. fitted out an expedition to Tahiti, under the command of the celebrated navigator Capt. James Cook. In order to make the angle as great as possible, and so increase the length of the chords, or paths of the planet across the sun, astronomers were sent to all the most favorable points of observation--St. Petersburg, Pekin, Lapland, California, etc. The result of these calculations fixed the solar parallax at 8.58″. This was considered accurate until lately, but has now ceased to have any value.

The next transits will happen,

December	8	1874.
"	6	1882.
June	7	2004.

The first transit ever seen was witnessed by Horrox, a young amateur astronomer residing near Liverpool. His calculations fixed upon Sunday, Nov. 24, 1639 (O. S.)

He however commenced his watch of the sun on Saturday preceding. On the following day he resumed his observation at sunrise. The hour for church arriving, *he repaired to service as usual.* Returning to his labor immediately afterward, he says: "At this time an opening in the clouds, which rendered the sun distinctly visible, seemed as if Divine Providence encouraged my aspirations; when—oh most gratifying spectacle! the object of so many earnest wishes—I perceived a new spot of perfectly round form that had just entered upon the left limb of the sun."

The transits of Mercury are more frequent; but owing to the nearness of the planet to the sun, they are of little value in determining the solar parallax.

The difficulty of determining the solar parallax accurately will be seen, when one is told that the correction from the old value of 8.58″ to the new one of 8.94″, is a change in the angle equal to that which the breadth of a human hair would make when seen at a distance of 125 feet. Yet this reduces the estimated distance of the sun from 95,293,000 miles, to 91,430,000 miles.

4. TO FIND THE LONGITUDE OF A PLACE.—(1.) *The solar method.*—If the sailor can see the sun, he watches it closely with his sextant; and when it ceases to rise any higher in the heavens it is *apparent* noon. By adding or subtracting the equation of time (as given in his almanac), he obtains the true or *mean noon*. He then compares the local time thus obtained, with the Greenwich time as kept by the ship's chronometer. The difference in time reduced to degrees, etc., gives the longitude.

(2.) *The lunar method.*—On account of the difficulty in obtaining a watch which will keep the exact Greenwich time through a long voyage, the moon is more generally relied upon than the chronometer.

The Nautical Almanac* is always published, for the benefit of sailors, three years in advance. It gives the distance of the moon from the principal fixed stars which lie along its path, at every hour in the night. The sailor has only to determine with his sextant the moon's distance from any fixed star, and then by referring to his almanac find the corresponding Greenwich time. By comparing this with the local time, and reducing the difference to degrees, etc., he obtains the longitude.

5. To FIND THE LATITUDE OF A PLACE.—(1.) By means of the sextant find the elevation of the pole above the horizon, and this gives the latitude directly. (2.) In the same manner, determine the height of the sun above the horizon at noon. The sun's declination for that day (as laid down in the almanac), added to or subtracted from this gives the height of the equinoctial above the horizon. Subtract this from 90°, and the remainder is the latitude.

---

* It is pleasant to notice that the astronomer can *predict* with the utmost precision. He announces that on such a year, month, day, hour, and second, a celestial body will occupy a certain position in the heavens. At the time indicated we point our telescope to the place, and at the instant, true beyond the accuracy of any timepiece, the orb sweeps into view! A prediction of the Nautical Almanac is received with as much confidence as if it were a fact contained in a book of history. "On the trackless ocean, this book is the mariner's trusted friend and counsellor; daily and nightly its revelations bring safety to ships in all parts of the world. It is something more than a mere book. It is an ever-present manifestation of the order and harmony of the universe."

6. To FIND THE CIRCUMFERENCE OF THE EARTH.—If the earth were a perfect sphere, it is obvious that degrees of latitude would be of the same length wherever measured on its surface. Each would be $\frac{1}{360}$ of the entire circumference. If, however, a person sets out from the equator, and travels along a meridian toward either pole, and when the polar star has risen in the heavens one degree above the horizon, he marks the spot, and then continues his journey, marking each degree in succession, he will find that the degrees are not of equal length, but increase gradually from the equator to the pole. If now the length of a degree be measured at different places, the rate of variation can be found, and then the *average* length be estimated. Measurements for this purpose have been made in Peru (almost exactly at the earth's equator), Lapland, England, France, India, Russia, etc. So great accuracy has been attained, that Airy and Bessel, who have solved the problem independently, differ in their estimate of the equatorial diameter but 77 yards, or only $\frac{44}{1000}$ of a mile.

7. To FIND THE RELATIVE SIZE OF THE PLANETS.—The volumes of two globes are proportional to the cubes of their like dimensions. The diameter of Mercury is 2,962 miles, and that of the earth 7,925; then,

The volume of Mercury : the volume of the earth :: $2962^3 : 7925^3$.

The same principle applied to the volume or bulk of the sun gives—

The bulk of the sun : bulk of earth :: $852584^3 : 7925^3$.

8. To FIND THE DIAMETER OF THE SUN.—(1.) A very simple method is to hold up a circular piece of paper before the eye at such a distance as to exactly hide the entire disk of the sun. Then we have the proportion,

As dist. of paper disk : dist. of sun's disk :: diam. of paper d. : diam. sun's d.

(2.) The apparent diameter of the sun, as seen from the earth, is about 32′: the apparent diameter of the earth, as seen from the sun, is twice the solar parallax, or 17.88″. Thence, the

Ap. diam. of earth : ap. diam. of sun :: real diam. of earth : real diam. of sun.

(3.) Knowing the apparent diameter of the sun, and its distance from the earth, the real diameter is found by Trigonometry. In figure 95, let S represent the earth, AB the radius of the sun, and ASB half the apparent diameter of the sun. We shall then have the proportion,

AS : AB :: radius : sin. 16′ (half mean diam. of sun).

By a similar method the diameters of the planets are obtained.

APPENDIX.

TABLE ILLUSTRATING KEPLER'S THIRD LAW. (CHAMBERS.)

In the first column are the relative distances of the planets from the sun; in the second, the periodic times of the planets; and in the third, the squares of the periodic times divided by the cubes of the mean distances. The decimal points are omitted in the third column for convenience of comparison. The want of *exact uniformity* is doubtless due to errors in the observations.

Vulcan ?	.143	19.7	132 716
Mercury.	.38710	87.969	133 421
Venus..........	.72333	224.701	133 413
Earth	1.	365.256	133 408
Mars...........	1.52369	686.979	133 410
Jupiter.........	5.20277	4,332.585	133 294
Saturn..........	9.53878	10,759.220	133 401
Uranus.........	19.18239	30,686.821	133 422
Neptune........	30.03680	60,126.710	133 405

Arago, speaking of Kepler's Laws, says: "These interesting laws, tested for every planet, have been found so perfectly exact, that we do not hesitate to infer the distances of the planets from the sun from the duration of their sidereal periods; and it is obvious that this method possesses considerable advantages in point of exactness."

Measurements of the Earth's Diameter.

	Airy.	Bessel.
Polar diameter.................	7899.17	7899.11
Equatorial diameter............	7925.64	7925.60
Compression	26.47	26.49

Table of

Elements of the Solar System.

Name.	Sign.	Diameter in miles.	Density. Earth as 1.	Density. Water as 1.	Greatest distance from the sun in miles.	Least distance from the sun in miles.	Mean distance from the sun in miles.	Sidereal period.	Synodic period.
								(Days.)	(Days.)
Sun......	☉	852,584	.25	1.43					
Moon....	☾	2,160	.63	3.57	251,947 (from earth.)	225,719 (from earth.)	238,833 (from earth.)	27.32	29.5
Mercury.	☿	2,962	1.24	7.03	42,665,560	28,119,716	35,392,638	87.96	115.8
Venus...	♀	7,510	.92	5.23	66,585,947	65,677,009	66,131,478	224.70	583.9
Earth....	⊕	7,925	1.	5.67	92,965,489	89,894,951	91,430,220	365.25	...
Mars.....	♂	4,920	.52	2.93	152,283,936	126,340,516	139,312,226	686.97	779.8
Jupiter..	♃	88,390	.22	1.23	498,603,768	452,782,530	475,693,149	4332.58	398.8
Saturn...	♄	72,904	.12	.68	921,105,027	823,164,139	872,134,583	10759.22	378.
Uranus...	♅	33,024	.18	.99	1,835,700,825	1,672,001,279	1,753,851,052	30686.82	369.7
Neptune.	♆	36,620	.17	.96	2,770,217,344	2,722,325,120	2,746,271,232	60126.71	367.5

ELEMENTS OF THE SOLAR SYSTEM—(*continued.*)

NAME.	Polar compres'n.	Axial rotation.	Light and heat of the sun. Earth as 1	Inclination of axis to the plane of ecliptic.	Force of Gravity. Earth as 1	Velocity in orbit. Miles per hour.	Mass (in tons).	Mean apparent diameter f'm earth.	Volume. Earth as 1.	Inclination of orbit to the ecliptic.
		D. H. M.								
Sun......	?	25 7 48		82° 30′	27.20		1,910,278,070,000,000,000,000,000,000	32′ 3″	1,245,126,361	
Moon....	?	27 7 43		88° 30′	.15	2,273	78,000,000,000,000,000,000	31′ 26″	.02034	5° 8′ 40″
		h. m. s.								
Mercury..	$\frac{1}{29}$	24 5 30	6.67	?	.46	105,330	393,000,000,000,000,000,000	8.7″	.052	7° 0′ 5″
Venus ...	slight.	23 21 23	1.91	?	.87	77,050	4,763,000,000,000,000,000,000	38.1″	.851	3° 23′ 29″
Earth....	$\frac{1}{299}$	23 56 4	1.	66° 32′	1.	65,533	6,069,000,000,000,000,000,000		1.	0°
Mars.....	$\frac{1}{55}$	24 37 23	.43	61° 18′	.30	53,090	750,000,000,000,000,000,000	17.3″	.239	1° 51′ 6″
Jupiter ..	$\frac{10}{168}$	9 55 21	.037	86° 55′	2.42	28,744	1,825,900,000,000,000,000,000,000	40.7″	1387.431	1° 18′ 52″
Saturn ...	$\frac{100}{922}$	10 29 17	.011	58° 41′	1.09	21,221	546,406,000,000,000,000,000,000	17.5″	746.898	2° 29′ 36″
Uranus...	?		.003	?	.73	14,963	76,721,000,000,000,000,000,000	3.9″	72.359	46′ 28″
Neptune.	?	?	.001	?	.79	11,958	101,720,000,000,000,000,000,000	2.8″	98.664	1° 46′ 59″

QUESTIONS.

THESE are the questions which the author has used in his own classes for review and examination. In the historical portion, he has required his pupils to write articles upon the character and life of the various persons named, gathering materials from every attainable source. He has also introduced whatever problems the class could master, taking topics from the article on Celestial Measurements and the various mathematical treatises.

INTRODUCTION.—Define Astronomy. Is the earth a planet? What is the difference in the appearance of a fixed star and a planet? What is the Milky Way?—HISTORY. What can you say of the antiquity of astronomy? How far back do the Chinese records extend? Name some astronomical phenomena they contain.

17. Why should the Chaldeans have become versed in this study? How ancient are their records? What discoveries did they make? What Grecian philosopher early acquired a reputation in this science? What other discovery did Thales make (Phil., p. 261)? What did he teach? What memorable eclipse did he predict? What were the names of two of his pupils? What did Anaximander teach?

18. Anaxagoras? What was his fate? In what century did Pythagoras live? What was his characteristic trait? Did he have any proof of his system? Explain his theory. How does it differ from ours? What strange views did he hold? What theory did Eudoxus advance?

19. What is the theory of the crystalline spheres? What has Hipparchus been styled? What addition did he make to astronomical knowledge? How many stars in our present catalogue (p. 228)? How did Egypt rank in science at an early day? What preparation did the Grecian philosophers make to fit themselves for teachers? How long did Pythagoras travel for this purpose? What can you say of the school at Alexandria? What great work did Ptolemy write there? What theory did he expound?

20. Was it original? What discovery did Eratosthenes make? Describe that method (p. 309). Show how the movements of the planets puzzled the ancients. What was the theory of "cycles and epicycles?"

21. Did the ancients believe in the reality of this cumbrous machinery? Did this theory possess any accuracy? Could they adapt it to explain any new motion?

22. What was the remark of Alphonso? When did astronomy cease to be cultivated as a science? In what century? Why did Cæsar import an astronomer? Why did he attempt to revise the Calendar? What change did he make (p. 295)? State something of the repute in which astrology was held.

23. Tell what you can of the system. What use did it subserve? What theory displaced the Ptolemaic? When? Was the system of Copernicus original? What credit is due him? Describe his idea of apparent motion. How did he apply this to the heavenly bodies?

24. What crudity did he retain? Who was Tycho Brahé? What was his theory? How did it differ from Ptolemy's and Copernicus's?

25. What good did he accomplish? Could he generalize his facts? Had he a telescope? How did Kepler differ from Brahé? What were the two prominent characteristics of Kepler? Give his three laws. Tell how he discovered the first. The second. The third (p. 313). Describe the ellipse. Define focus, perihelion, and aphelion. What remarkable statement did Kepler make?

29. When did Galileo live? What discoveries did he make in Natural Philosophy? In Astronomy? What advantage did he have over his predecessors?

30. Give an account of his observations on the moon. On Jupiter's moons.

31. Why did this settle the controversy between the Ptolemaic and the Copernican system? How were Galileo's discoveries received? Give some of Sizzi's ponderous arguments.

33. Who discovered the law of gravitation? Repeat it. How was this idea suggested? What familiar laws of motion aided Newton? How did he apply these to the motion of the moon? Repeat the story of his patient triumph.

35. What is the celestial sphere? Give the two illustrations which show its vast distance from the earth.

36. Why can we not see the stars by day, as by night? What portion of the sphere is visible to us? Name the three systems of circles.

37–41. Name and define (1) the principal circle, (2) the

secondary circles, (3) the points, and (4) the measurements of each system. Define, especially, because in common use, zenith, nadir, azimuth, altitude, equinoctial, right ascension, declination, equinox, ecliptic, colure, and solstice. What is N or S in the heavens? What is the Zodiac?

42. How wide? How ancient? How divided? Give the names and signs. State the meaning of each (p. 229).

II. The Solar System.

Of what is the solar system composed? Describe how we are to picture it to ourselves.

The Sun.—Its sign. Its distance from us? Illustrate.

47. How are celestial distances measured? To what is the sun's light equal? To how many full moons? Its heat? Illustrate. What proportion of the sun's heat reaches the earth?

48–50. Its apparent size? How does this vary? Its dimensions—(1) diameter—illustrate, (2) volume, (3) mass, (4) weight, (5) density. How large did Pythagoras think the sun is? Tell something about the force of gravity in the sun. How much would you weigh if carried to its surface? (This can be calculated from the table in Appendix.) How does the sun appear to the naked eye? How can we see the spots? What were formerly the views of astronomers with regard to the sun's face? When were the spots discovered?

52. Tell something about the number of the spots. Their location. Size.

53. Describe the parts of which they are composed. The motion of the spots.

54–5. How do they change in form as they pass across the disk? What does this prove? What is the length of a solar axial revolution? Explain a sidereal and a synodic revolution.

56–7. Why do not the spots move in straight lines? Show how they curve. Tell what you can about the irregular movements of the spots. Tell how suddenly they change.

58. What can you say about their periodicity? Who discovered this? Is there any connection between the solar spots and the aurora? Tell the influence of the planets on the spots. Explain.

59. Do the spots affect the fruitfulness of the season? Does the temperature of the spots differ from that of the rest of the sun? Are they depressions in the sun? How much darker are they than the adjacent surface?

60. Is the sun brighter than the Drummond light? *Ans.* The sun gives out as much light as one hundred and forty-six

lime-lights would do, if each were as large as the sun and were burning all over. What are the faculæ? Describe the mottled appearance of the sun.

61. What is the shape of the bright masses? What is a pore?

62. Describe the constitution of the sun according to Wilson's theory. How are the spots produced? The faculæ?

63. The penumbra? The umbra?

64. What is Kirchhoff's theory? How are the spots produced? The umbra? The penumbra? Upon what does this theory depend (p. 286)? What is the cause of the heat of the sun? Will the heat ever cease?*

THE PLANETS.—Name the six characteristics common to all the planets.

67. Compare the two groups of the major planets.

68. Draw an ellipse, and name the various parts. Define the ecliptic.† The ascending node. The descending node. Line of the nodes. Longitude of the node. Tell what you can with regard to the comparative size of the planets.

71. What is a conjunction? Name the earliest that are recorded.

72. Tell what you can concerning the planets being inhabited.

74. What about the conditions of life on the different planets? What are the two divisions of the planets?

75. What causes the apparently irregular movements of the planets? Define heliocentric and geocentric place. Illustrate. In what part of the sky is an inferior planet always seen? Define inferior and superior conjunction.

76. Greatest elongation. Quadrature. Why is a star at one time "evening" and at another "morning star?"

77. What is a transit? Explain the retrograde motion of an inferior planet. (This motion, it will be remembered, was one that sorely puzzled the ancients.)

---

*If we accept the Nebular hypothesis (p. 283), we must suppose that the heat is produced by the condensation of the nebulous matter and consequent chemical changes. The sun is radiating its heat constantly, and, after a time, will go out, in turn, as the earth and all the planets have before it.

† Lockyer beautifully says: "We may imagine the earth floating around the sun on a boundless ocean, both sun and earth being half immersed in it. This level, this plane, the plane of the ecliptic (because all eclipses occur in it), is used by astronomers as we use the sea-level. We say a mountain is so far above the level of the sea. The astronomer says a star is so high above the level of the ecliptic."

78. Describe the phases of an inferior planet. Why does an inferior planet have phases? Define gibbous.

79. Explain the opposition and conjunction of a superior planet. Its retrograde motion. Must a superior planet always be seen in the same part of the sky as the sun?

80. Which retrogrades more, a near or a distant planet? Define a sidereal and a synodic revolution of an inferior and a superior planet, and tell what you can about each. In what case would there be no difference between a sidereal and a synodic revolution? Why is a planet invisible when in conjunction?

82. When is a planet evening, and when morning star? Tell what you can about the supposed discovery of a planet interior to Mercury.

83. MERCURY.—Definition and sign? Describe the appearance of Mercury, and where seen.

84. What was the opinion of the ancients? The astrologists? Chemists? Why is it difficult to see it? When can we see it best?

85. What is the peculiarity of its orbit? Its distance from the sun? Velocity? Length of its day? Year? Difference between its sidereal and synodic revolution? why? Its distance from the earth?

86. Show why its greatest and least distances vary so much. What is its diameter? Volume? Density? Force of gravity? Specific gravity? How much would you weigh on Mercury? Describe its seasons. (If the pupil does not understand pretty well the subject of the terrestrial seasons, it would be well here to read carefully page 110, et seq.)

88. Its temperature? Appearance of the sun? Has it any moon? What is the appearance of the planet through a telescope? What do these phases prove? What do we know of its mountains and valleys?

89. VENUS.—Definition and sign? Ancient names? Appearance to us?

90. When brightest? Can Venus be seen by day? Illustrate.

91. Describe the orbit. What is the distance of Venus from the sun? Velocity? Length of the year? Day? Difference between the sidereal and synodic revolution? Distance from the earth?

92. How does the apparent size vary? When is Venus the brightest? What is the diameter? Volume? Density?

93. Force of gravity? Does the force of gravity increase

or decrease with the mass or volume of the body? Describe the seasons.

94. Describe the telescopic appearance. Who discovered the phases of Venus? What was Copernicus's idea?

95. What proof have we of an atmosphere? Of clouds? Has Venus any moon?

96. EARTH.—Sign? What is the appearance of the earth from the other planets? Do we, then, live on a star? Is it probable that the earth was always dark and dull as it now seems to us?* How does the size of the earth compare with that of the other planets? Form of the earth? Exact diameter? Is the equator a perfect circle?

98. Circumference? Density? Weight? What can you say of its inequalities? How do you prove the rotundity of the earth?

99. Why can we see further from the top of a hill than from its base? Why is the horizon a circle?

100. Give some illustrations of apparent motion.

101. Explain the cause of the rising and setting of the sun and stars. Who first explained it in this manner? What do you say of its simplicity?

102. Cause of day and night? Do all places on the earth revolve with equal velocity? Illustrate. At what rate do we move?

103. Why do we not perceive our motion? What would be the effect if the earth were to stop?

104–5. Is there any danger of this catastrophe? Draw the figure, and show how the stars move daily through unequal orbits and with unequal velocities. Describe the appearance of the stars at the N. Pole.

106. At the Equator. S. Pole. Describe the path of the earth about the sun. Define eccentricity. Is this stable?

107. Do we see the same stars at different seasons of the year? Why not? If we should watch from 6 P. M. to 6 A. M., what portion of the sphere could we see? What do we mean by the yearly motion of the sun among the stars? How can we see it?

109. What is the cause? What is the ecliptic? Why so called? What are the equinoxes? What do we understand

---

* Probably not. The earth was doubtless once a glowing star, like the sun. Its crust is only the ashes and cinders of that fearful conflagration. The rocks are all burnt bodies. The atmosphere is only the gas left over after the fuel was all consumed. Every organic object has been rescued by plants and the sunbeam from the grasp of oxygen.

when we see in the almanac "the earth is in Aries?" "The sun is in Sagittarius?"

110. How many apparent motions has the sun? Name them, and give the cause and effects of each. Has the sun any real motions (pp. 54 and 224)? Describe the apparent motion of the sun, N. and S. How is it that the sun in summer shines on the north side of some houses both at rising and setting, but in winter never does? Define the obliquity of the ecliptic. The parallelism of the earth's axis. What do you say of its permanence?

112. Why will a top stand while spinning, but will fall as soon as it ceases? Show how the rays of the sun strike the various parts of the earth at different angles at the same time. Show how the angles vary at different times. Is the sun really hotter in summer than in winter? Why does it seem to be?

113. Explain the cause of equal day and night at the Equinoxes. Why are our days and nights of unequal length at all other times? Why do they vary at different seasons of the year? How do the seasons, &c., in the N. Temperate Zone compare with those in the S. Temperate Zone? Describe the yearly path of the earth about the sun—(1) at the summer solstice; (2) at the autumnal equinox; (3) at the winter solstice; (4) at the vernal equinox; (5) the yearly path finished back to the starting-point. Is the division of the earth's surface into zones an artificial or a natural distinction? Who invented it?

117. How much nearer are we to the sun in the winter? Why is it not the warmest at that time? How is it in the South Temperate Zone? When do the extremes of heat and cold occur? Why not exactly at the solstices?

118. Why is summer longer than winter? Does the earth move with the same velocity in all parts of its orbit? Describe the curious appearance of the sun at the North Pole. In Greenland, at what part of the year will the midnight sun be seen due north? What is the length of the days and nights at the Equator?

119. Describe the results if the axis of the earth were perpendicular to the ecliptic.

120. If the equator were perpendicular to the ecliptic. Define precession of the equinoxes. Who discovered this? At what rate does this movement proceed? What is the amount at present?

121. What are the results? What star was formerly the Pole star?

123. Explain the cause of precession.

125. How does the spinning of a top illustrate this subject?

126. What is Nutation? Cause? How does the moon's influence compare with that of the sun?

127. What is the real path of the N. Pole through the heavens? Is the obliquity of the ecliptic invariable? What is the limit? What is the effect of this variation?

128. Are the solstices and equinoxes stationary? What is the result of this change on the seasons? When will the cycle be completed? When is the sun in perigee?

129. What do you say of the provisions made to secure permanence, so that slight changes themselves prevent greater changes?

130. What is refraction? Its effect?

131. How does it vary?

132. Are the sun and moon ever where they seem to be? Is the real day longer or shorter than the apparent one? Why do the sun and moon appear flattened when near the horizon? Why not when they are high in the heavens? Why do they appear smaller in the latter case?

133. What causes the hazy appearance of the heavenly bodies near the horizon? What is the cause of twilight? How long does it last? Is it the same at all seasons of the year?

134. At all parts of the earth? Where is it longest? Shortest? What is diffused light? What would be the effect if the atmosphere did not act in this way? Is there really any sky in the heavens? Cause of the appearance?

135. What is aberration of light? Illustrate. Give two reasons why we never see the sun where it really is.

137. The general effect of aberration? Define parallax. Illustrate.

138. Define true and apparent place. How does parallax vary? What is the practical importance of this subject (p. 300, et seq.)?

139. Define horizontal parallax. What is the sun's horizontal parallax? What is the annual parallax?

THE MOON.—Signs? Describe its orbit.

140. Its distance from the earth? Illustrate. Difference between its sidereal and synodic revolution?

141. What is the real path of the moon? (Imagine a pencil fastened to the spoke of a wheel, and the wheel rolled by the side of a wall on which the pencil is constantly marking.) How often does it turn on its axis? What is the moon's diameter? Volume? How does its apparent size vary? Why does it appear larger than it really is?

142. Why does the crescent moon appear larger than the dark body of the moon? When ought the moon to appear the largest? Do all persons think the moon of the same apparent size? Explain the three librations of the moon.

143. How does moonlight compare with sunlight? Is there any heat in moonlight? Why is it generally clear at full moon? Does the centre of gravity in the moon coincide with that of magnitude? Has the moon any atmosphere? What proof have we of this?—*Ans.* (1) We see but slightly if any appearance of twilight in the moon. (2) When the moon passes between us and a star, it does not refract the light of the star, so that the atmosphere cannot be sufficient to support more than $\frac{3}{100}$ of an inch of the mercurial column.

144. How does the earth appear from the moon? What is the earth-shine? How is it caused? What is it called in England? Describe the path of the moon around the earth, and the consequent phases. Why is new moon seen in the west and full moon in the east? Why can we sometimes see the moon in the west after the sun rises, and in the east before the sun sets?

147. Length of a lunar month? What do we mean by the moon's running high or low? Cause? Use?

148. What is harvest moon? Hunter's moon? Cause?

149. What are nodes? How much is the moon's orbit inclined to the ecliptic—our ideal sea-level? What is an occultation? Use?

150. Describe the seasons, heat, &c., on the moon.

152. Telescopic appearance of the moon? Are the mountains the light or dark portions? What can you say about them? The gray plains? The rills? The craters? What are the peculiar features, then, of the lunar landscapes? Are the lunar volcanoes extinct?

ECLIPSES.—When can an eclipse of the sun occur? Show how a solar eclipse may be total, partial, or annular.

156. Define umbra. Penumbra. Central eclipse. State the general principles of a solar eclipse.

158. What curious phenomena attend a total eclipse?

159. Describe the effect of a total eclipse?

160. What curious custom prevails among the Hindoos? What is the Saros? Cause?

161. Is it now of any value? What is the metonic cycle? Explain its use.

162. What is the golden number? Cause of a lunar eclipse? Draw the figure and describe it. Why are lunar eclipses seen oftener than solar ones?

163. What is the earliest account of an eclipse? How were eclipses formerly regarded?

164. What story is told of Columbus?

THE TIDES.*—Define ebb. Flow. How often does the tide happen? Explain the cause.

166. Why does the tide occur fifty minutes later each day? Why is there a tide on the side opposite the moon? The sun is much larger than the moon; why does it not produce the larger tide? Why is not the tide felt out at sea?

167. What is spring-tide? Neap-tide? Causes? Why does the tide differ so much in various localities? Tell about the height of the tide at different points.

168. Why is there no tide on a lake?

MARS.—Definition and sign?

169. Describe its appearance. When is it brightest? Its distance from the sun? Velocity? Day? Year?

170. Distance from the earth? Peculiarity of its orbit? Diameter? Volume? Density? Mass? Force of gravity? Figure? Describe its seasons.

171. Has it any atmosphere? Moon? Appearance of our earth? Telescopic features. (The land and sea features have been so well decided that they have been named, and a Mars's globe made.)

172. Cause of its ruddy color? What are the snow-zones? Can we watch the change of its seasons?

MINOR PLANETS (ASTEROIDS).—Give Bode's law. Tell how the first of these planets was discovered. How many are now known?—*Ans.* There are (Sept. 21,1870) 112. Are they probably all discovered?

174. Describe these "pocket planets." Are they all found within the Zodiac? What is their origin?—*Ans.* According to the Nebular hypothesis, the ring of matter broke up into numberless small bodies instead of aggregating into one large planet. Give some of the names and signs.

JUPITER.—Definition and sign? Describe its appearance. Ancient views. Describe its orbit. What is its distance from the sun? Velocity? (1869. ♃ is in ♈). Day? Year? Distance from the earth?

177. Diameter? Volume? Density? Centrifugal force?

* As the tidal wave does not move as rapidly as the earth does, the water has an apparent backward motion. It has been suggested that this *acts as a break* on the earth's diurnal revolution. It has been shown that the moon's true place can be best calculated if we suppose that the sidereal day is shortening, by tidal action, at the rate of $\frac{1}{66}$ of a second in 2,500 years.

Force of gravity? Figure? Describe its seasons. Upon what does the change of seasons in any planet depend?

178. The appearance of the sky? The telescopic features? Are Jupiter's moons visible to the naked eye?

179. How named? What is their size? What space do they occupy?

180. Describe the eclipse of the moons.

181. Define immersion, emersion, and transit. How rapidly do the satellites revolve? What can you say of the frequency of eclipses on Jupiter? Describe the belts. Why are they parallel to its equator?

182. How was the velocity of light discovered?

SATURN.—Definition and sign? Describe its appearance. How rapidly does it move through the sky? (1869. ♄ is in ♏). Its distance from the sun? Peculiarity of its orbit?

184. Velocity? Year? Day? Distance from the earth? Diameter? Volume? Density? Force of gravity? Describe its seasons.

185. Has it any atmosphere? Who discovered the rings of Saturn? Describe them.

186. Are they stationary? Explain their phases.

187. Describe Saturn's belts.

188. Describe Saturn's moons. The scenery on Saturn.

URANUS.—Definition and sign? How was it discovered? Tell of its previous discovery by Le Monnier. Is Uranus visible to the naked eye? (1869. ♅ is in ♋). Distance from the sun? Year? Diameter? Density?

191. Describe its seasons. Telescopic features. Satellites Peculiarity of its moons.

NEPTUNE.—Definition and sign? Appearance in the sky? Give an account of its wonderful discovery.

193. What is its distance from the sun? Year? Velocity? Diameter? Volume? Density? Do we know anything of the seasons? Why not? Intensity of the light?

194. Appearance of the heavens? What are the telescopic features? Has Neptune any moon? What advantage have the Neptunian astronomers?

METEORS, AEROLITES, AND SHOOTING-STARS.—Define an aerolite. A shooting-star. A meteor. Give some account of the fall of meteors (aerolites).

197. What elements are found in aerolites? How can an aerolite be distinguished? Give an account of wonderful meteors.

198. Of shooting-stars.

199. Describe the showers of 1799 and 1833.

200. The shower of 1866. At what intervals did these showers occur? Why was not the shower of 1866 seen in this country? *Ans.* Our side of the earth was not turned toward the meteors.

201. What is the average number of meteors and shooting-stars daily? Why do we not see more of them?

202. In what months are they most abundant? Describe the origin of meteors and shooting-stars. What is their velocity? What causes the light? The explosion often heard? What is said of a companion to our moon?

203. What is the theory of meteoric rings? What is their shape? How do these account for the showers at regular in tervals?

204. What is the period of the November ring? Why is the August shower so uniform, while the November one is only periodic?

205. What is the relation between meteors and comets? What do you mean by the radiant point? What effect do meteors have on the weather?

206. What is their height? Weight?

COMETS.—How were they looked upon by the ancients? Illustrate. Define the term comet. What is the tail? The nucleus? The head? The coma? Does each comet necessarily possess all these parts? How would a mere round, fleecy mass be known to be a comet? What mistake did Herschel make in looking, as he supposed, at one of this kind (p. 189)?

208. Where do comets appear? In what direction do they move? How does a comet look when first seen? Upon what does the time of greatest brilliancy depend? What do you say of the number of the comets? What was Kepler's remark?

209. Why do we not see them oftener? Where did Seneca see one? Describe the orbits of comets. Which class has been calculated? Which classes never return?

210. Describe the difficulty of calculating a comet's orbit.

211. Name the periods of some. What has been the distance from the sun of some noted comets? Velocity?

212. What do you say of the density of a comet? Illustrate. Is there any danger of our running against a comet?

213. Do comets shine by their own or by reflected light? Tell what you can of their variation in form and dimensions.

214. Give some account of the comets of 1811, 1835, and 1843. For what is Encke's comet noted? What is its period? Give some description of Donati's comet.

ZODIACAL LIGHT.—Where can this be seen? What is its appearance? Where is it brightest? What is its origin?

III.—The Sidereal System.

Tell something of the appearance of the heavens at Neptune's distance from the sun—our starting-point? Do we ever see the stars? What do we see, then?

222. Which star is nearest the earth? What is its parallax? Its distance? What is Prof. Airy's remark?

223. How long would it take light to reach the nearest star? How would the earth's orbit appear at that distance? Our sun? How long does it take for the light of the smaller stars to reach the earth? What can you say of the motion of the fixed stars? Illustrate.

224. What proof have we that the stars are suns? ("If Sirius shines as brightly as our sun, at its distance, it must be three thousand times larger."—Lockyer.) That our sun is only a small star? Describe the motion of the solar system. What is the centre? How many stars can we see with the naked eye? With a telescope? Have all the stars been discovered?

226. What is the cause of the twinkling of the stars? Do the stars twinkle in tropical regions? Why not? What do you say of the magnitude of the stars? Name four points of difference between a planet and a fixed star.

227. What do you mean by a star of the first magnitude? How many are there? Of the second magnitude? How many sizes may one see with the naked eye? With a telescope? What is the cause of the difference in the brightness of the stars? What can you say of the names of the stars?

228. What can you say with regard to the division of the stars into constellations? Is there any real likeness to the mythological figures? Name any figure which seems to you well founded.

229. Are the boundaries distinct? Who invented the system? Give the meaning of the signs of the Zodiac and their origin.

230. Explain why the signs and constellations of the Zodiac do not agree.

231. What causes the appearance of the constellations? Would they appear as they now do, if we should go out into space among them?

232. Are the present forms permanent? State the value of the stars in practical life.

233. What were the views of the ancients with regard to the stars?

234. Describe the division of the stars into three zones, and name them.

THE CONSTELLATIONS.—The questions on these are uniformly: (1) *description*, (2) *principal stars*, and (3) *mythological history*. They need not therefore be repeated with each constellation.—What are the pointers? Does Polaris mark the exact position of the North Pole? How many times per day is Polaris on the meridian of any place? Explain how this applies in navigation or surveying. State how the amount of the variation from the true north will change through the ages. What star will ultimately become the pole-star? What curious facts are stated concerning the Pyramids? What do you say of the distance of Polaris? How may latitude be calculated by means of Polaris?

DOUBLE STARS, ETC.—Does any star appear double to the naked eye? How many have been found by the use of the telescope? What is an optical double star? Are all double stars of this class? Describe the revolution of a binary system. What other combinations have been discovered? Their periods?

266. Orbits? Mass? Are these companion stars as close to each other as they seem? What can you say of the colored stars? Do their colors ever change? Which colors would indicate the hottest star?

267. What is the probable effect in a system having colored suns? What are variable stars? Describe the changes of Algol.

268. Of Mira. What is the cause? What are temporary stars? Describe the one seen in Cassiopeia.

269. The one in Corona Borealis, in 1866. What are lost stars?

270. Can you give any explanation? Of what did the star of 1866 consist? Are these stars destroyed? Is the process of creation now complete?

271. What are star clusters? Name several.

272. Is such a grouping a mere optical effect? Are they probably as closely compacted as they seem to be? What are nebulæ? How do they differ from clusters? Is it probable that all nebulæ will be resolved into clusters? What has spectrum analysis proved some of the nebulæ to be?

273. Are they suns? Where are they most abundant? What can you say about their distances? Into how many classes are they divided? Describe and illustrate the elliptic nebulæ. What is said of the distance of the great nebula in Andromeda? The number of stars it contains? Describe

the annular nebulæ. What is said of the "ring universe" in Lyra?

276. Its diameter? Describe the spiral nebula in Canes Venatici. Describe the planetary nebulæ. What is said of the number and size of these "island universes?"

277. Describe the fantastic appearance of the irregular nebulæ. What are nebulous stars? What is the cause?

278. What is said of the size of the one in Cygnus? What are variable nebulæ?

279. Give instances. What is said of double nebulæ? Is anything definite known with regard to them? What are the Magellanic clouds?

280. Describe the Milky-way. What can you say of the number of stars in the Galaxy? Are the stars uniformly distributed?

281. What is Herschel's theory of the constitution of the universe? If this theory be true, what is our sun?

282. Give an account of the Nebular hypothesis. What is said of Saturn's rings? May they ultimately disappear?

284. What is spectrum analysis? Name the three kinds of spectra.

285. What colored rays will a flame absorb? Describe the spectroscope.

286. What are Fraunhofer's lines? What is known of the constitution of the sun? What proof have we that iron exists in the sun?

287. What elements have been found in the sun? What proof have we that the stars are suns? What can you say of the similarity existing between the stars and our earth?

288. What has been discovered with regard to the constitution of the Nebulæ? Of their relative brightness?

TIME.—What two methods of measuring time? What is a sidereal day?

289. What are astronomical clocks? Tell how they are used. Why do astronomers use sidereal time? What is a solar day? What causes the difference between a sidereal and a solar day? To how much time is a degree of space equal?

290. Which is taken as the unit, the solar or the sidereal day? How long is a solar day? A sidereal day? A solar day equals how many sidereal hours? A sidereal day equals how many solar hours? Describe mean solar time. What is apparent noon? Mean noon? The equation of time? When is this greatest? When least?

291. When do mean and apparent time coincide? Can a watch keep apparent time? How may apparent time be kept?

How can it be changed into mean time? Tell how to erect a sun-dial. When will a sidereal and a mean-time clock coincide? A mean-time clock and the sun-dial?

292. Give the two reasons why the solar days are of unequal length.

294. What is the civil day? Who invented the present division? Describe the customs of various nations. What is the origin of the names of the days?* What is the sidereal year? The mean solar year? What causes the difference?

295. What is the anomalistic year? How did the ancients find the length of the year? What error did they make? What was the result? Give an account of the Julian calendar. The Gregorian calendar. What is the meaning of the terms O. S. and N. S.? What country now uses O. S.? When was the change adopted in England? How was it received? How could a child be eight years old before a return of its birthday?

297. When do the Jews begin their year? Why does our year begin January 1st? Show how the earth is our timepiece. What influence has Jupiter's moons on the cotton trade?

CELESTIAL MEASUREMENTS.—These problems are to be used throughout the study. They require no questions but the formal statement of the problem requiring solution.

---

* It is said that the Egyptians named the seven days from the seven celestial bodies then known. The order was continued by the Romans. Tuesday they called *Dies Martis*; Wednesday, *Dies Mercurii*; Thursday, *Dies Jovis*; Friday, *Dies Veneris*. In the Saxon mythology, Tius, Woden, Thor, and Friga are equivalent to Mars, Mercury, Jupiter, and Venus. Hence we see the origin of our English names.

GUIDE TO THE CONSTELLATIONS.

THE following is a description of the appearance of the heavens on or about the first day of each month in the year.

January. (7 P. M.)—*In the North*, Cassiopeia and Perseus are above Polaris, Cepheus and Draco west, Ursa Minor below, and Ursa Major below and to the east. *In the East*, Cancer is just rising, Canis Minor (Procyon) has just risen. *Along the Ecliptic*, Gemini is well up, then Taurus, Aries reaches to the meridian, next Pisces, Aquarius (letter Y) and Capricornus just setting. *In the Southeast*, Orion and the Hare are well up. *In the South*, Cetus swims his huge bulk far to the east and west. *In the Southwest* is Piscis Australis (Fomalhaut). *North of the Ecliptic* the Triangles are nearly in the zenith, Perseus is just east, below is Auriga, Andromeda lies just west of the meridian, and Pegasus is midway, while Delphinus (the Dolphin, Job's Coffin), Aquila (Altair), and Lyra (Vega) are fast sinking to the western horizon.

February. (7 P. M.)—*In the North*, Ursa Major lies east of Polaris, Ursa Minor and Draco below, Cepheus west, Cassiopeia above and to the west. *In the East*, Regulus and Cor Hydræ are just rising. *Along the Ecliptic*, Leo (Regulus, the sickle) just rising, Cancer well up, Gemini midway, Taurus on the meridian, Aries (the scalene triangle) past, Pisces next, and lastly Aquarius just setting. *In the Southeast*, Canis Minor, Canis Major (Sirius), and Orion are conspicuous. *In the Southwest*, Cetus covers nearly the whole sky. *North of the Ecliptic*, Perseus is on the meridian, while Auriga is a little east of it; west of Perseus is Andromeda, while the great square of Pegasus is fast approaching the horizon.

March. (7 P. M.)—*In the North*, Ursa Major lies east of Polaris, Draco and Ursa Minor below, Cepheus below and to the west, and Cassiopeia west. *In the East*, Cor Caroli (the Greyhound) is well up, and Coma Berenices is rising. *Along the Ecliptic*, Leo is fully risen, next Cancer, Gemini reaches to the meridian, Taurus is past, Aries midway, and lastly Pisces is just beginning to set. *In the Southeast*, Cor

Hydræ, Canis Minor and Canis Major are conspicuous. *In the South,* Orion blazes brilliantly. *In the Southwest,* Cetus is hiding below the horizon. *North of the Ecliptic,* Auriga is in the zenith; west are Perseus and Andromeda, while Pegasus is just beginning to sink out of sight.

April. (7 P. M.)—*In the North,* Ursa Major is above and to the east of Polaris; opposite and to the west is Perseus, Draco below and to the east, Cepheus below and to the west, Cassiopeia west. *In the East,* Bootes (Arcturus) not quite fully risen. *Along the Ecliptic,* Virgo (Spica) rising, Leo midway, Cancer reaches to the meridian, Gemini past, next Taurus, then Aries, and lastly Pisces just setting. *In the Southeast* is the Crater (the Cup), and Hydra stretches its long neck to the meridian. *In the South,* Canis Minor. *In the Southwest,* Sirius and Orion. *North of the Ecliptic,* and in the northeast, are Coma Berenices and Cor Caroli; above Gemini and Taurus is Auriga, while Andromeda is just setting in the northwest.

May. (8 P. M.)—*In the North,* Ursa Major is above Polaris, Ursa Minor and Draco east, Cepheus and Cassiopeia below, and Perseus west. *In the East,* Lyra is just rising, and Hercules is just up. *Along the Ecliptic,* Libra is just rising, Virgo is midway, Leo is on the meridian, Cancer is past, next Gemini, and lastly Taurus just setting. *In the South,* stretching east and west of the meridian, is Hydra, with the Crater and Corvus a little east. *In the Southwest,* is Cor Hydræ, Canis Major, and Canis Minor, while Orion is just setting in the west. *North of the Ecliptic,* in the east, above Hercules, are Corona Borealis (The Northern Crown), Bootes (Arcturus), Coma Berenices, and Cor Caroli, which stretch nearly to the meridian. *In the Northwest,* above Taurus and Perseus, is Auriga.

June. (8 P. M.)—*In the North,* Ursa Major is above Polaris, Draco and Ursa Minor to the east, Cepheus below and to the east, and Cassiopeia directly below. *In the East,* Cygnus and Aquila are just rising, Lyra and Taurus Poniatowskii are well up. *Along the Ecliptic,* Scorpio is rising, Libra is midway, Virgo on the meridian, Leo past, Cancer midway, Gemini next, and Taurus just setting. *In the South* are Corvus and the Crater, a little past the meridian. *In the Southwest* is Cor Hydræ, and in the west Canis Minor approaching the horizon. *North of the Ecliptic,* in the east, above Scorpio, is Hercules; then Corona and Bootes, and near the meridian, Cor Caroli and Coma Berenices. *In the Northwest* is Auriga, just coming to the horizon.

July. (9 P. M.)—*In the North,* Draco and Ursa Minor

above Polaris, Ursa Major west, Cepheus east, and Cassiopeia below to the east. *In the East*, the Dolphin (Job's Coffin) is now well up, Cygnus is almost midway to the meridian, and Lyra is still higher. *Along the Ecliptic*, Capricornus is rising, Sagittarius (the Archer) is next, Scorpio, with its long tail swinging along the horizon, is directly south, Libra is past the meridian, Virgo midway, and Leo has almost reached the horizon. *In the Southwest*, the Crater is setting, and Corvus is just above. *North of the Ecliptic*, above Scorpio and east of the meridian, are Serpentarius, Hercules, and Taurus Poniatowskii; Corona is almost on the meridian, to the west of which lie Boötes, Cor Caroli, and Coma Berenices.

August. (9 P. M.)—*In the North*, Draco and Ursa Minor are above Polaris, Cepheus above and to the east, Cassiopeia east, and Ursa Major west. *In the Northeast*, Perseus is just rising, while south of it Andromeda and Pegasus are fairly up. *Along the Ecliptic*, Aquarius is risen, next Capricornus, Sagittarius reaches to the meridian, Scorpio is just past, Libra next, and Virgo (Spica) just touches the horizon. *North of the Ecliptic*, Taurus Poniatowskii is on and Lyra is just east of the meridian; the Swan and Dolphin are east of Lyra, Serpentarius and Hercules are above Scorpio, and just west of the meridian; thence west are Corona and Boötes, while far in the northwest are Coma Berenices and Cor Caroli.

September. (8 P. M.)—Draco is above and to the west of Polaris, Cepheus above and to the east, Cassiopeia east, Ursa Major is below and to the west. *In the Northeast*, Perseus is just rising. *In the East*, Andromeda is fairly up, Pegasus is nearly midway to the meridian. *Along the Ecliptic*, Pisces is just rising, next Aquarius, Capricornus in the southwest, Sagittarius on the meridian in the south, next Scorpio in the southwest, Libra, and lastly Virgo just setting. *North of the Ecliptic*, Lyra is on the meridian, Cygnus, the Dolphin, and Aquila just to the east, while to the west are Taurus Poniatowskii and Serpentarius; north of these latter are Hercules, Corona, Boötes, Cor Caroli, and Coma Berenices.

October. (7 P. M.)—*In the North*, Cepheus and Draco are above Polaris, Ursa Minor west, Cassiopeia east, and Ursa Major below and west. *In the Northeast*, Perseus is fairly risen. *In the East*, Andromeda is nearly midway to the zenith. *Along the Ecliptic*, Aries is just rising, Pisces well up, Aquarius and Capricornus in the southeast, Sagittarius in the south, Scorpio far down in the southwest, and Libra just setting. *North of the Ecliptic*, Cygnus and Aquila are on the meridian, the Dolphin just east of it, and far south; Lyra is west of the meridian, Taurus Poniatowskii lower down and to the south, Serpentarius is just above Scorpio; next, in line with it and Polaris, is Hercules; Corona and Boötes are toward the northwest, where Coma Berenices is just setting.

November. (7 P. M.)—*In the North*, Ursa Major is below Polaris, Ursa Minor and Draco are to the west, Cepheus above, and Cassiopeia above and to the east. *In the Northeast*, Auriga is just rising, and Perseus is above, nearly midway to the meridian. *Along the Ecliptic*, Taurus is just rising, next Aries and Pisces; Aquarius is on the meridian, south; then Capricornus, and lastly Sagittarius, in the southwest. *North of the Ecliptic*, Pegasus and Andromeda lie east of the meridian, the Swan, Dolphin, Eagle, Taurus Poniatowskii, and Lyra west. *In the Northwest* are Hercules and Corona.

December. (7 P. M.)—*In the North*, Cassiopeia is above Polaris, Cepheus above and to the west, Perseus above and to the east, Draco west, and Ursa Major below. *In the Northeast*, below Perseus, is Auriga. *In the East*, Orion is rising. *Along the Ecliptic*, Gemini is just rising, Taurus is nearly midway, next Aries, Pisces is on the meridian, then Aquarius, and lastly Capricornus, far in the southwest. *In the South*, east of the meridian, is Cetus, and west is Fomalhaut. *North of the Ecliptic*, Andromeda is nearly on the meridian, and Pegasus west of it; Cygnus, Delphinus, Lyra, and Aquila are about midway, while Taurus Poniatowskii is just sinking to the horizon. *In the Northwest*, Hercules is just setting.

NOTE.—It should be borne in mind that a month makes a variation of about two hours (30°) in the rise of a star: hence, in the foregoing "Guide," the "January Sky" of 9 P. M. would be about the same as the "February Sky" of 7 P. M.; the "January Sky" of 11 P. M. would be about the same as the "March Sky" of 7 P. M., &c. In this way the "Guide" may be used for any hour in the night. The pupil will see that in the "Guide" the prominent figures and stars in each constellation are given in parentheses. Examples: the "Y" in Aquarius, the "scalene triangle" in Aries, "Job's coffin" in the Dolphin, "Procyon" in Canis Minor, &c. These aid in identifying the constellation.

INDEX.

www.ingramcontent.com/pod-product-compliance
Lightning Source LLC
LaVergne TN
LVHW020210110826
845151LV00003B/668

* 9 7 8 1 4 2 5 5 3 4 5 1 6 *